WILEY SERIES OF PRACTICAL CONSTRUCTION GUIDES

M.D. Morris, P.E., Series Editor

Cortland A. Collier and Don A. Halperin
CONSTRUCTION FUNDING: WHERE THE MONEY COMES FROM, Second Edition

Ben C. Gerwick, Jr.
CONSTRUCTION OF OFFSHORE STRUCTURES

Richard G. Ahlvin and Vernon Allen Smoots
CONSTRUCTION GUIDE FOR SOILS AND FOUNDATIONS, Second Edition

B. Austin Barry
CONSTRUCTION MEASUREMENTS, Second Edition

Harold J. Rosen and Tom Heineman
CONSTRUCTION SPECIFICATIONS WRITING: PRINCIPLES AND PROCEDURES, Third Edition

Leo Diamant and C.R. Tumblin
CONSTRUCTION COST ESTIMATES, Second Edition

Thomas C. Schleifer
CONSTRUCTION CONTRACTORS' SURVIVAL GUIDE

A. C. Houlsby
CONSTRUCTION AND DESIGN OF CEMENT GROUTING: A GUIDE TO GROUTING IN ROCK FOUNDATIONS

David A. Day and Neal B. H. Benjamin
CONSTRUCTION EQUIPMENT GUIDE, Second Edition

Terry T. McFadden and F. Lawrence Bennett
CONSTRUCTION IN COLD REGIONS: A GUIDE FOR PLANNERS, ENGINEERS, CONTRACTORS, AND MANAGERS

S. Peter Volpe and Peter J. Volpe
CONSTRUCTION BUSINESS MANAGEMENT

Julian R. Panek and John Philip Cook
CONSTRUCTION SEALANTS AND ADHESIVES, Third Edition

J. Patrick Powers
CONSTRUCTION DEWATERING: NEW METHODS AND APPLICATIONS, Second Edition

William R. Park and Wayne B. Chapin
CONSTRUCTION OF PRESTRESSED CONCRETE STRUCTURES, Second Edition

Ellis J. Krinitzsky, James P. Gould, and Peter H. Edinger
FUNDAMENTALS OF EARTHQUAKE RESISTANT CONSTRUCTION

Ben C. Gerwick, Jr.
CONSTRUCTION OF PRESTRESSED CONCRETE STRUCTURES, Second Edition

John J.P. Krol
CONSTRUCTION CONTRACT LAW

CONSTRUCTION OF FILLS

CONSTRUCTION OF FILLS

Second Edition

EDWARD J. MONAHAN
Bloomfield, New Jersey

JOHN WILEY & SONS, INC.

New York • Chichester • Brisbane • Toronto • Singapore

This text is printed on acid-free paper.

Library of Congress Cataloging in Publication Data:

Monahan, Edward J.
Construction of fills / Edward J. Monahan. — 2nd ed.
p. cm. — (Wiley series of practical construction guides)
Rev. ed. of: Construction of and on compacted fills. 1986.
"A Wiley-Interscience publication."
Includes bibliograhpical references (p.) and index.
ISBN 0-471-58523-8 (alk. paper)
1. Building. 2. Fills (Earthwork) 3. Soil stabilization.
I. Monahan, Edward J. Construction of and on compacted fills.
II. Title. III. Series.
TH153.M56 1994
624.1′5—dc20 93-29806

Printed in the United States of America

10 9 8 7 6 5 4 3 2 1

To

My wife, Mary-Jean, for her constant support in general, and especially for supporting me with regard to important career decisions coinciding with the period of the writing of this book.

and to

David McKay Greer, my early "teacher" in the professional practice of geotechnical engineering, with the hope that he will get away from drilled piers for a while, so that he can complete *Reminiscences of an Arkansas Traveler.*

A foolish consistency is the hobgoblin of little minds.

—RALPH WALDO EMERSON

A respect for complexity is the beginning of wisdom.

—DEAN RUSK

Everything should be made as simple as possible, but not simpler.

—ALBERT EINSTEIN

SERIES PREFACE

Congratulations! You've just bought a profit-making tool that is inexpensive and requires no maintenance, no overhead, and no amortization. Actually, it will increase in value for you each time you use this volume in the Wiley Series of Practical Construction Guides. This book should contribute toward getting your project done under budget, ahead of schedule, and out of court.

For nearly a quarter of a century, over 50 books on various aspects of construction and contracting have appeared in this series. If one is still valid, it is "updated" to stay on the cutting edge. If it ceases to serve it goes out of print. Thus you get the most advanced construction practice and technology information available from experts who use it on the job.

The Associated General Contractors of American (AGC) statistician advises that the construction industry now represents close to 10% of the gross national product (GNP), some 410 billion dollars worth per year. Therefore, simple, off-the-shelf books won't work. The construction industry is unique in that it is the only one where the factory goes out to the buyer at the point of sale. The constructor takes more than the normal risk in operating a needed service business.

Until the advent of the series, various single books (many by professors), magazine articles, and vendors' literature constituted the total source of information for builders. To fill this need, this series has provided solid usable information and data for and by working constructors. This has increased the contractors' earning capacity while giving the owner a better product. Profit is not a dirty word. The Wiley Series of Practical Construction Guides is dedicated to that cause.

M. D. Morris, P.E.

Ithaca, New York
November 1989

PREFACE TO THE SECOND EDITION

The title of the second edition has been changed to reflect the fact that most of the additions deal with fills that are not compacted.

The major new work is Chapter 9, Septic Systems, which includes descriptions of the sequential steps of this important technology: Site Investigation and Evaluation, and Design and Construction, followed by three aspects of attention to existing systems: Renovation and Repair, Maintenance, and System Management. Emphasis is given to Site Investigation and Evaluation and to Maintenance because it is my feeling that these aspects of the technology are not usually given the attention they deserve. I also feel that proper attention to these neglected aspects can yield the best return for the least expenditure of time, cost, and effort, both in terms of return on investment and with respect to the protection of the environment.

Extensive new material has been added in Sections 5.2 and 5.3 concerning the burgeoning use of artificial fills throughout the world (foam plastics, Elastizell, Geocell, Solite), and the use of waste materials as fills (shredded tires, wood chips). The use of such fills has potential multiple benefits, including the constructive use of materials that normally present problems of disposal, the consequent mitigation of pollution problems (such as tire storage, fire hazards, and the related potential for air and groundwater pollution), and the avoidance of dealing with the growing problem of contaminated soil fills, resulting in the growing scarcity of "clean fills," especially in urban areas. A proposal is described that hopefully will encourage greater attention to research and development on the thermal properties of fills toward the eventual use of fills for the conservation of energy.

New material dealing with the correlation of standard and modified Proctor densities, and a method for determining numerical values of bearing capacities for Proctor densities are included (Sections 4.3 and 4.4). Case studies are used to augment the descriptions.

In Section 11.3, Seminars and Lectures, I have added my recommendation for a two-day compaction seminar, given by F. C. Budinger, P.E. I participated in one of these seminars (as a luncheon speaker), and can attest to the exceptional value of the presentation. I have communicated with Mr. Budinger, and learned that he offers the seminar on both a scheduled individual-registration basis, and on an in-house "by arrangement" basis. Thus, large soils engineering companies, earthwork contractors, or equipment manufacturers may wish to invite him to present his seminar to their staff. With this book and his seminar, I guarantee a full and complete education in compaction technology.

Those who have read the first edition know that I value plain, clear language (with occasional humor thrown in), sometimes to the point of being quite blunt, so let me establish that tone here in the preface. As is inferred by many of the case studies in the book, I have been almost constantly amazed at the number of large earthwork contractors who know little or nothing about compaction technology. It is perhaps understandable that a guy with a pickup truck and a backhoe might be ignorant of soils compaction technology, but a contractor who has many many millions of dollars in capital equipment and who regularly bids on major jobs is quite another matter. Not to invest the relatively tiny amount of time and money toward the "compaction education" of his or her staff has never made any sense to me. For example, one of the things that I recall from Budinger's seminar was his emphasis on the importance of moisture control in the field, asserting that "the most important pieces of equipment [are] the sprinkler truck and the related mixing equipment needed to adjust field moisture of placed fills." Most contractors, at least in my experience, would regard such requirements as an annoyance at best, and nonsense at worst, and would resist the imposition of such demands, or go along with great reluctance, all because they do not appreciate the importance of correct field moisture toward easy and efficient field compaction. I submit that, if they take the time to learn, their profits will increase, and field work will go a lot more smoothly and amicably, simply because the contractor's representative will *understand and agree to* the required adjustments. Knowledge is power *and* profits.

Finally, I offer my own services as a lecturer on aspects of septic system technology, with an unusual twist. Since retiring from engineering education in 1984, I have gotten into storytellilng and, believe it or not, I am developing a story about a privy builder. A personal friend, Bayonne Jim Kontje, sent me a little book, entitled *The Specialist*, written by a fellow named Chic Sale. At first, I thought it was an obscure effort by an equally obscure author, but later learned it was a best-seller some time in the 1920s (600,000 copies!), and that

Chic Sale was a comedian who also had a flair for writing. Needless to say, the book has some pretty funny descriptive passages. The result: *The Specialist.* Accordingly, Chapter 9, Septic Systems, is appropriately dedicated to the memory of Chic Sale, and to Lem Putt, the privy builder "specialist" of the story (who, incidentally, was a real person). For those interested in the seminar or the lecture, addresses are listed in Section 11.3.

I hope you enjoy the book, and maybe even some day, the story..

EDWARD J. MONAHAN

Pocono Pines, Pennsylvania
April 1993

PREFACE TO THE FIRST EDITION

This book is in the Wiley Series of Practical Construction Guides. It deals with compacted fills and is written principally to be helpful to the construction contractor. To serve that purpose, however, it is necessary to provide practical and useful information to *all* nonspecialists who traditionally are involved with or affect the planning, plans, specifications, and execution of earthwork construction. Such nonspecialists include the architect, the structural engineer, and the fill inspector, whose activities have a strong impact on the contractor's performance and profits.

Recent spectacular failures of man-made earthen structures have been extremely costly in terms of both dollars and lives. For example, on Sunday, November 6, 1977, at approximately 1:30 A.M., The Kelly Barnes Dam near Taccoa, Georgia, failed. The resulting flood took 39 lives. As a direct result of the Taccoa Dam disaster, a national dam inspection program was initiated.

The principal cause of the Taccoa failure has been judged by geotechnical experts to have been piping, a progressive interior deterioration of the earth structure associated with seepage forces of sufficient magnitude to dislodge and slowly transmit particles along with flowing water.

As imposing as the losses associated with nationally reported disasters are, I am convinced that these losses pale in comparison to damage caused by the countless thousands of defective earth structures—or fills—that are constructed annually. The number of cracked sidewalks, pavements, floor slabs, walls and columns, and basement walls must be prodigious. Improperly backfilled trenches resulting in traffic hazards and damaged utilities can be both costly and dangerous. The costs escalate significantly when litigation becomes a factor, especially where personal injury or deaths are involved.

A peculiarity of earthwork construction is that many nonspecialists who are routinely involved frequently do not consult with geotechnical specialists for advice regarding filling operations. I believe the reasons for this are (1) a belief that filling operations are simple and straightforward and do not require the attention (and expense) of specialists and (2) a preoccupation with the (eventual) visible aspects of the structure such as architectural considerations and structural design.

The principal reasons for this undesirable situation are (1) the neglect of the topics of soil compaction and fill control in *required* undergraduate courses in civil engineering and civil engineering technology and (2) that little information seems to be available to *nonspecialists* involved in earthwork aspects of engineered construction.

I hope that this book will correct or at least alleviate the foregoing problems.

EDWARD J. MONAHAN

Bloomfield, New Jersey
March 1986

ACKNOWLEDGMENTS

I would like to acknowledge the help of Tom Bellatty, P.E., who got me in trouble by recommending me to Walter Waidelich, P.E., to do a paper for the 72nd Annual Meeting of the Transportation Research Board, held in Washington, D.C., in January 1993. "Wally" chaired two sessions, one on artificial fills and one on waste fills. The new material in Sections 5.2 and 5.3 would not have been nearly as extensive if I had not picked some brains in Washington and later communicated with the fine engineers I met there: Messrs. Humphrey, Edil and Boscher, Ahmad and Lovell, Upton and Machan, Allen and Killian, and Negussey. Special thanks also to Professor Edil for sending a copy of the excellent research report on shredded tires, a report that was not a part of the presentations in Washington. The willingness of these gentlemen to share information is commendable and appreciated. Their papers and affiliations are cited in the pertinent sections and in the list of references.

I am indebted to Joseph P. Conway, P.E., Director and Assistant Commissioner, Department of Environmental Protection, City of New York, for following up on his promise to send me information about the successful uses of "oil-eating bacteria" for sewer problems in New York City and elsewhere. As described in Section 9.5.4, I believe there may be significant potential for improving the performance of septic systems by the use of this new technology.

Special thanks also to James C. Converse, of the University of Wisconsin, for sending the fine reports and bibliography on septic systems that represent the excellent work of investigators at Wisconsin.

Thanks to John Schmertmann, of Schmertmann and Crapps, Gainesville, FL, for supplying the information about the flat dilatometer test (Chapter 9, Glossary).

Thanks also go to Professor Bob Stoll of Columbia University for supplying information about the career and writings of his predecessor and former colleague, the late Donald M. Burmister.

I thank Ed Mott for his valuable input regarding septic system performance (Chapter 9).

I am greatly indebted to Ralph Peck, for a number of reasons. First, he sent information to me regarding the work on foam plastic fills done by the Norwegians. Second, he has encouraged me greatly with his reviews of both the first edition and my proposal for this second edition. I believe I have complied with all of his constructive suggestions for this edition, and I am sure that his generous comments about the first edition were the main force behind the book's success and acceptance by the profession.

I also thank Dr. Peck for the inspiration that he has provided by his exemplary career and writings. I think he will be pleased to learn that for over 20 years of my teaching career, the very first assignment to my undergraduate students (numbering in the thousands, I would estimate) was his beautifully written article, "Art and Science in Subsurface Engineering," *Geotechnique*, March 1963. To those who have not read it, please do. As all of my former students know, I have always asserted that the ability to write and speak clearly is at least as important as technical competence. Since Dr. Peck's article contains great advice on the "observational method" so vital to the practice of geotechnical engineering, and is written so well and so cogently, it was, I thought, a perfect vehicle to "launch" young engineering students on their careers, whether or not they eventually opted for geotechnical engineering as their specialty. I know for a fact that a number of my former students in professional practice (many of them now principals) got a great running start from that fine article.

I would like to thank my editor, Dan Sayre, for allowing me my style, and my Production Editor, Millie Torres. Reviewers of the first edition used terms such as "conversational," "practical," "interesting," "straightforward," "jaunty," "lively," "humorous," "entertaining," and "full of wisdom" (Ralph Peck made the latter two comments). Above all else, I value the most common assessment: *Readability*. That is what I set out to do, especially since I come from a background where that quality is rare: academia. (Sorry about that, guys.)

Last, but certainly not least, I thank my very greatest supporter and friend, my wife, Mary-Jean, who not only put up with my "holing up" in the Poconos to concentrate on the writing of the manuscript, but encouraged me to do so, while she tended to things at home and commuted to the Poconos during our temporary "bi-statal existence." I started during the Blizzard of 1993, but right now, it's 70° and sunny . . . I'm goin' fishin'.

CONTENTS

1. Introduction **1**

1.1 Purpose and Scope 1
1.2 Importance and Nature of Earthwork 2
1.3 Important Definitions 3
1.4 The Role of Index and Engineering Properties 7
1.5 Glossary 10

2. Avoiding Costly Blunders **12**

2.1 The Practical Value of Knowledge of the Historical Development of Soil Compaction 12
2.2 Early Empirical Approaches 12
2.3 Rational Approach 13
2.4 Standard Proctor Density 13
2.5 Modified Proctor Density 14
2.6 Load-Bearing Fills/Building Codes 15
2.7 Summary 16
2.8 Glossary 16

3. Basics of Soil Compaction Curves: Laboratory Procedures **17**

3.1 Compaction Defined 17
3.2 Spectrum of Soil Types 19
3.3 Curve Locations and Shapes and Their Practical Meaning: Moisture and Energy Effects 21

3.4 ASTM Compaction Requirements 22
3.5 Summary 24
3.6 Glossary 25

4. Major Problems in Compacted Fill Technology: Proposed Solutions 26

4.1 Standard-Modified Ignorance 26
4.2 The Ninety-five Percent Fixation 29
4.3 Correlations of Standard and Modified Proctor Densities 38
4.4 Bearing Capacities of Proctor Densities 38
4.5 Changing Borrow 40
4.6 Problems Evolving from Traditional Practice 43
4.7 Cost and Time Pressures: A Summary 45
4.8 Glossary 46

5. Applied Research and Development 47

5.1 Effects of Specific Index Property Variations 47
5.2 Artificial Fills 52
5.3 Waste Materials as Fills 73
5.4 Effects of Mechanical Laboratory Compactors 80
5.5 Density Gradients 81
5.6 Geostick Correlations 81
5.7 Densities of Uncompacted Fills 81
5.8 Percent Compaction Specifications for Clay Fills 83
5.9 Summary 84
5.10 Glossary 84

6. Fills and Fill Compaction 86

6.1 Strength, Stability, and Imperviousness: Contrasting Requirements 86
6.2 Potential Problems with Earth Structures 87
6.3 Controlled and Uncontrolled Fills 101
6.4 Nonstandard and Special Fills 101
6.5 Compactors and Lift Thicknesses 103
6.6 Energy and Moisture Control 110
6.7 Glossary 112

7. Compaction Specifications **114**

7.1 Typical Specifications 115
7.2 Implementation and Enforcement 126
7.3 Nontechnical Aspects of Specifications 129
7.4 Specification and Project Evaluation 129
7.5 Glossary 132

8. Fill Control Procedures—Inspection **134**

8.1 Field-Density Testing 134
8.2 The Compleat Field Inspector 156
8.3 Case Histories 166
8.4 Glossary 189

9. Septic Systems **190**

9.1 Background Technology 191
9.2 Site Investigation and Evaluation 195
9.3 Design and Construction 206
9.4 Renovation/Repair of Marginally Performing Systems 229
9.5 Maintenance 230
9.6 Septic System Management 240
9.7 Glossary 241

10. Techonomics **242**

10.1 Engineering Design 242
10.2 Legal Costs 243

11. Appendices **247**

11.1 Very Large Jobs 247
11.2 Annotated References for Further Study 249
11.3 Seminars and Lectures 252

References **253**

Index **259**

CONSTRUCTION OF FILLS

CHAPTER 1

Introduction

1.1 PURPOSE AND SCOPE

The purpose of this book is to present helpful information on soil compaction and fill control to nonspecialists. The information should also be useful to academic colleagues, and especially so if I am successful in convincing them of the importance of a greater emphasis in their courses, especially those that are heavily attended by nonspecialists. Because of the focus noted, established on the basis of 33 years of observations in both the classroom and in consulting practice, I assume no prior knowledge of soil compaction, or, for that matter, soil mechanics. A careful study of Chapters 1, 2, and 3 is recommended for a better appreciation and understanding of that which follows.

Chapter 4 presents the major problems in fill control, and solutions are proposed for each problem. Some are strictly technical in nature, while others deal with much more sensitive matters ranging from professional ethics to human relations. Observations made about the latter are bound to be controversial. However, I am convinced that what is said needs to be said, openly and forthrightly, in order to have any chance of correcting certain practices detrimental to good engineered construction.

To provide credibility and validation to the assertions made regarding major problems, case studies are used. In all cases, names, locales, and other possibly embarrassing details are omitted.

Chapter 5 describes some of my unpublished research results and suggests further research efforts. I believe investigations can be pursued at many levels—funded research, undergraduate and graduate laboratory courses, and undergraduate and graduate projects and theses. I hope that geotechnical consultants adopt some of the suggestions and improve and extend certain procedures to augment their recommendations for, and supervision of, filling operations. In some cases, for example, the suggestions regarding a compaction data book (for dealing with the problems of *changing borrow*), their ability to accumulate large amounts of data makes it feasible to develop, refine, and validate methods rapidly.

Chapter 6, Fills and Fill Compaction, deals in detail with the technological aspects of all types of compacted fills, and includes a section on potential problems with earth structures, with an emphasis on earthen dams. The intent is to provide information helpful to those involved with the National Dam Inspection Program, thus serving the dual purpose of improving capabilities in the inspection and remediation of completed earth structures and in the full range of activities needed for design, inspection, and construction of new earthworks. Chapter 7 complements Chapter 6 with a description of all details relating to compaction specifications.

Chapter 8 (plus some aspects of Chapter 7, notably Specification Evaluation, Section 7.4) is intended to serve as a manual for fill control procedures: It is written expressly for the typical inexperienced young geotechnical engineer or engineering technologist.

As a departure from typical format, chapter glossaries are provided at the end of each chapter to explain certain terms used in the chapter that could not be explained fully within the text without adversely affecting readability. I suggest that the reader review each chapter's glossary before reading the chapter of interest.

1.2 IMPORTANCE AND NATURE OF EARTHWORK

The importance of soil compaction and fill control can be emphasized by one simple declaration: Almost no significant engineered construction occurs without the movement of soil from one place to another. Furthermore, it should be the nature of good engineered construction that parties become involved in earthwork operations in the following sequence: the geotechnical engineer, the architect and structural engineer, the fill inspector, and the construction contractor. A most important nonspecialist is, of course, the owner or client. Unfortunately, as noted in the preface, geotechnical engineers are too often left out of the sequence. All too often, they are called in (late) to correct a bad situation or to act as expert witnesses when it is already too late for correction.

In the ideal sequence, the geotechnical engineer explores and evaluates the subsurface conditions through a logical process of analyzing soil (or rock) *index properties*, and then, as judgment dictates, determines appropriate *engineering properties* through laboratory and/or field testing. With a knowledge of the loading conditions that are to be imposed by the proposed structure, the engineer prepares recommendations for foundation type, methods of foundation construction, and allowable bearing capacities at particular foundation levels. Such recommendations are broadly based upon determining safe loading intensities on foundation elements (e.g., footings) of various sizes that will not result in either a *bearing capacity failure* or *unacceptable settlements.* More often than not, the recommendations will entail excavation and filling operations, requiring recommendations for *quality* (texture) and *condition* (compacted density) of the fill. Assuming compliance with specifications, the fill is then judged to have a certain allowable bearing capacity.

The structural engineer is one of the primary recipients of the geotechnical engineer's report. In addition to the obvious responsibility of structural design, it is his typical responsibility to prepare working drawings, including foundation drawings. The preparation of construction specifications becomes a natural extension of this work, typically as notes on the drawings and (for large projects) separate, additional documents. In the usual situation, the structural engineer will need to consult architectural drawings and geotechnical reports and communicate with the architect and geotechnical engineer when questions develop requiring their attention and expertise.

Construction engineers and contractors are, of course, the "doers." They bring to fruition the studies, concepts, designs, drawings, and specifications of the architect, the geotechnical engineer, and the structural engineer, thus completing the sequence of operations of engineered construction.

1.3 IMPORTANT DEFINITIONS

Soil Compaction (Field) The reduction of void spaces (densification) of lifts of fill by the direct application of load, impact, and/or vibration, usually with a suitable type of compaction equipment. Lift thicknesses vary from several inches (for clays and silts) to perhaps 2 ft for free-draining fills (sands and gravels). (*Note:* One should not confuse soil compaction with consolidation, the long-term reduction of void ratio of a natural soil, usually saturated, thick deposits of soft clays or silts beneath the water table. This is usually accomplished by the application of static surface loading (called surcharges), resulting in the slow drainage of porewater from the subsurface stratum. Thus, one principal difference is that compaction is direct and immediate.)

Soil Compaction (Laboratory) The compaction of a small but representative soil sample, obtained from the field, in a steel mold of standard size. The soil is compacted usually in layers, commonly by dropping a hammer of specified weight through a specified distance a specified number of times. The energy of such compaction is chosen to simulate that of field compaction, usually with a roller judged suitable for the conditions and soil type. The moisture content is varied for a series of filled molds, thus generating a *compaction curve* for the soil. A typical curve is shown in Figure 1.1.

Granular, Cohesionless Soils Gravels, sands, and "clean" silts (those possessing no plasticity).

Cohesive, Plastic Soils Clays, clay–silt mixtures, organic soils.

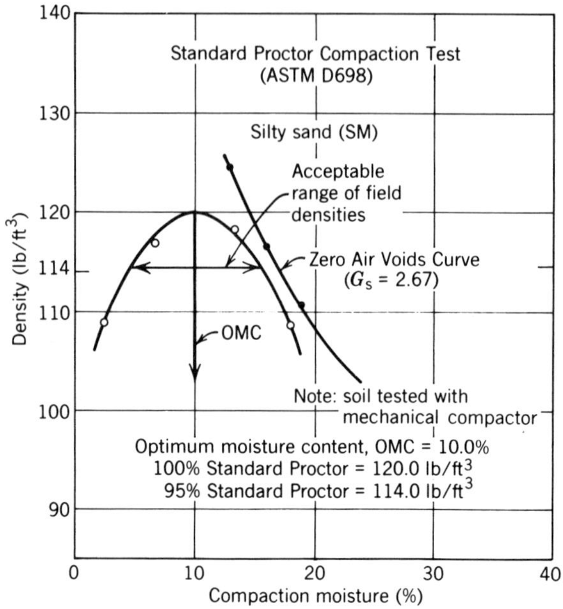

Figure 1.1. *A typical compaction curve.*

Drainage Quality Designations

1. *Free-draining soils*—gravels, coarser sands, and mixtures thereof.
2. *Marginally draining soils*—finer sands, clean silts, and mixtures thereof.
3. *Impervious soils*—clays, clay–silt mixtures. (Note that no soil is impervious, so the term is used in a relative rather than absolute sense.)

Gemicoss James Kilpatrick, one of my favorite columnists and lingophiles, suggests that if there is a legitimate need for a new word, invent one. Herewith, *gemicoss.* It is defined as any soil that is a combination of soil types in such proportions as to raise questions pertaining to contrasting engineering properties, for example, cohesionless vis-à-vis cohesive, plastic vis-à-vis nonplastic, free-draining vis-à-vis marginally draining. It may also be helpful to tell you how I invented the word: As a child, I learned that all (most?) English words required one or more vowels. Thus G is for gravel, C is for clay, SS is for sand and silt. E, O, and I, of course, are needed vowels. The letter I might also represent inorganics, the soil minerals. In the Unified Soil Classification System, O is used to represent organics, and M to indicate silts. EM also stands for Ed Monahan.

There are many examples illustrating the need for the concept of the gemicoss. Often a chart or a formula has a principal limitation of being applicable to a certain soil type, that is, sand or clay. Blow counts, for example, are used for determining allowable bearing capacities *for sands.* For clays, one often uses the unconfined compressive strength. But what does one do with a sand–clay mixture? Or a sand–silt–clay mixture? There is no simple answer, other than to say that good engineering judgment must be employed.

Essentially Granular (or Cohesive) Soils Texture that predominates in the context of dictating the overall behavior of the soil. This terminology is a natural extension of the notion of a gemicoss. After carefully evaluating the texture (sizes *and* plasticity) of the soil, one may be forced to decide whether it is essentially granular or essentially cohesive in order to justify the use of a particular design chart or formula.

For example, if a soil is composed of 60% highly plastic clay (by weight), with 40% sand and gravel, the soil would be an ECS, an essentially cohesive soil, since the sand and gravel particles are, for all practical purposes, merely isolated or "suspended" in a clay matrix, and their presence is essentially irrelevant to the overall engineering behavior of the soil. Note that this example does not fit the classification of a gemicoss, because the conclusion (ECS) is fairly evident. A gemicoss might be 40% medium plastic clay, 60% sand and gravel. Such a mixture raises questions pertaining to contrasting engineering properties.

As a contrasting example, if a soil is 75% gravel and sand, with 25% clay of high plasticity, it would be rated an essentially granular soil (EGS), inasmuch as it is reasonable to assume that the preponderance of granular particles are in contact. The significant percentage of highly plastic clay would act as "binder."

Thus, in summary, the simple auxilliary classification system that I propose includes five categories: granular, cohesionless soils; essentially granular soils; essentially cohesive soils; cohesive (plastic) soils; and the gemicoss.

Fill The soil that is selected for use at the site. It may be from a site excavation or "imported" from elsewhere. In the latter case, the fill is termed *borrow*.

Fill Quality Indices Texture: grain sizes and grain size distribution of cohesionless soils, or plasticity (usually plasticity index, or PI) of cohesive soils.

Fill Condition Density (pounds per cubic foot, pcf) or relative density (percent), a measure of potential settlement. For clay soils, potential *expansion* is often an important additional factor to consider.

Permanent Stability A concept referring to the fundamental question of whether the fill (or any soil) will remain stable under present *and future* conditions that may reasonably be expected to be imposed during the economic life of the structure. Thus, a very dense, free-draining soil, which is to be permanently contained (laterally), is the most "permanently stable" soil, for it cannot settle significantly, it is not susceptible to seepage pressures, nor will it expand when wetted. Conversely, a cohesive (clay) soil can settle unacceptably if not compacted sufficiently but can expand detrimentally upon (future) wetting if overcompacted. The pressures associated with such expansion can be substantial, causing damage to highway pavements and even heavy structures. Clays and silts are also susceptible to disturbance by seepage pressures (piping, uplift) because of their relative imperviousness. Silts are also susceptible to frost action.

Simulation of Field Conditions This is an important concept concerning the rationale that should be used for all decisions relating to laboratory testing. Thus, what one does to the soil sample in the laboratory should, in all practical respects, simulate what will be done to the soil in the field. As one of the more obvious examples, the method and energy of compaction in the laboratory compaction test should approximate that of construction rollers used in the field.

Index Properties Indicates the general nature of the subsurface problems that must be confronted. Such index properties are (or can be) obtained routinely and at modest expense in a typical subsurface investigation. Figure 1.2 shows, in simple flowchart form, how index properties should be used. Note that large jobs, requiring the expenditure of larger amounts of money, would justify the expenditure of larger amounts for field or laboratory testing to determine engineering properties needed for design. Engineering properties are defined as those measuring stress, strain, and strain rate on representative soil samples that enable computation of specific numerical values of bearing capacity and settlement. Thus, path ABC provides a design that is more reliable than path AC, but at considerably more cost. Path AC, for small jobs, where extensive testing is not economically justified, utilizes index properties directly for design purposes, usually by entering a chart to select allowable bearing capacities, rather than the more expensive ABC route. Of considerable practical importance, moreover, is the fact that index properties, if obtained and properly evaluated, serve as the basis for a rational testing program (step B). In summary, index properties serve a twofold purpose and are indispensable tools in geotechnical engineering.

Blow Count, Density, and Relative Density Illustrative of the foregoing description of index and engineering properties are the definitions of blow count, density, and relative density, particularly as related to compaction and fill control and the stability and bearing capacity of granular, or essentially granular, soils.

When a driller is sampling a soil that is expected to be an essentially granular soil, a sampling device is driven into the soil and a blow count N (blows per foot) is obtained. Clearly, the blow count (with any appropriate corrections) is an indicator of density; the higher the blow count, the higher the density.

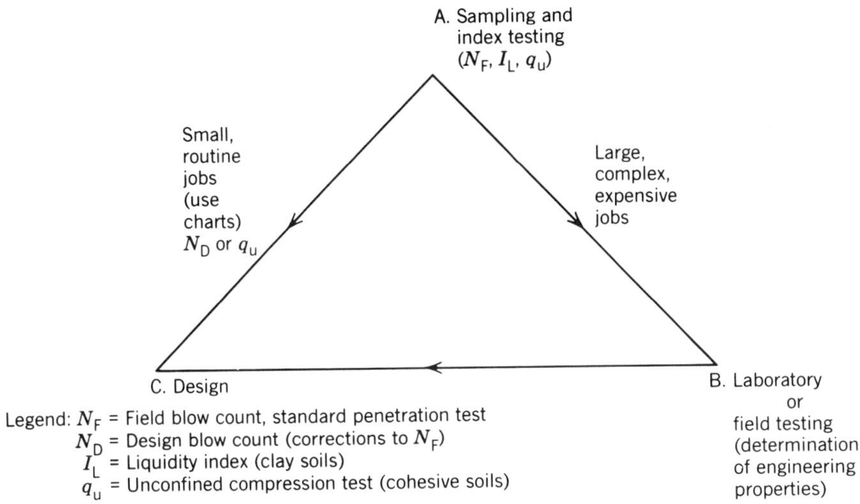

Figure 1.2. *Design flowchart.*

Relative density, a laboratory-determined *engineering property*, is a percentage of the practically obtainable maximum density for the particular soil. Thus, a soil of 100% relative density would have a negligible potential for future settlement. Furthermore, given the thickness of the soil, a knowledge of the relative density would allow for the computation of a specific, numerical settlement estimate. A judgment could then be made as to whether the computed settlement would be tolerable.

In this example, the blow count is an indicator (or index property) signifying the *general* condition of the soil (e.g., very loose, loose, medium-dense, dense, very dense), whereas the relative density (involving the added expense of laboratory testing) is an engineering property.

With respect to fills, knowledge of the relative density and the compacted density—which the designer can control by appropriate specifications—permits similar computations for settlement.

Rather than list and describe all index and engineering properties in this introductory chapter, terms will be defined where first used or will be included in the chapter glossaries.

1.4 THE ROLE OF INDEX AND ENGINEERING PROPERTIES

If the texture and blow counts of a soil being considered as potential borrow are determined from exploratory drilling and sampling operations, what practical information can be derived from such index data?

1.4.1 Case 1

Case 1 involves a well-graded gravelly sand, with approximately 25% fines of significant plasticity. (PI on minus-40 fraction is determined to be 30. Liquid limit (LL) is 65.) Blow count range, 35–60.

DEFINITIONS AND DISCUSSION. Fines are all soils passing the No. 200 sieve, thus silts or clays. If fines are judged by simple and quick field tests to be significantly plastic (clayey rather than silty), Atterberg Limits tests are ordered on that portion of the soil passing the No. 40 sieve: the minus-40 fraction. Result is PI = 30; LL = 65. Such fines would be rated "high plasticity," CH in the Unified Soil Classification System. In summary, this soil is a dense to very dense, well-graded, essentially granular soil, but containing cohesive clay binder. In a real situation, the locale of the potential borrow area would, of course, be known. Thus, soil maps, geologic maps, and perhaps personal knowledge of the area (geologic, topographic, and land-use, for example) would be available to augment the data from the exploratory program. The soil described is typical of a glacial till, soil deposited directly by a glacier in a mechanical fashion as opposed to alluvial deposition (by say, glacial meltwater). This accounts for the wide range of sizes, as contrasted to uniform sizes associated with the sorting action of flowing water. Thus, one could reasonably expect to encounter boulders in the area. Confirmation could be obtained by reconnaissance, particularly by inspection of road cuts in the vicinity.

Before proceeding with a listing of the practical and potentially very valuable information that can be gleaned from the foregoing, I feel it is necessary to add some commentary here regarding a very important subject: soil (and rock) descriptions as they pertain to formal classification systems. There are several such systems. The most commonly used, in the sense of broadest acceptance in the United States, are the Unified system and the AASHTO system (the American Association of State Highway and Transportation Officials). The best system, in my opinion, is the Burmister system, developed by Donald Burmister of Columbia University. It is outside the scope of this book to present details concerning classification systems or to editorialize extensively about the merits or shortcomings of each. Suffice it to say that the Burmister system, used extensively by many of my colleagues in professional practice in the New York/New Jersey area, forces the person using the system to be precise in estimating and then describing the texture of a soil sample, but not unreasonably or unnecessarily so. With concentration and practice, one can rather quickly develop the skill of estimating the proportions of constituents (gravel, sand, silt, clay) to within about 15% accuracy, the gradation of each granular constituent (coarse, medium, fine), and the approximate plasticity of any fines (silts, clays) present. Using only visual and tactile senses (sight, feel, smell—even taste, with one soil scientist I know!) formal, unambiguous descriptions can be written in accordance with the explicit rules governing the system. More information about this system is presented in Chapter 9.

The description given for Case 1, although it will serve its purpose here, is not in accordance with any formal system. Thus, in professional practice, it would be of little value, since "well-graded," "essentially granular," and "containing clay binder" have no well-defined meaning to other potential users. A soil classification system, then, is seen as a universally agreed upon language, wherein all words and symbols used have specific definitions. When people describe soils without adopting a recognized system, no one, save themselves, can be sure of what they mean. Unfortunately, it is all too common to see soil descriptions in practice, generally written by one of our nonspecialists, that are ambiguous, vague, and often virtually useless because of usage of terminology not belonging to any recognized system.

For a brief yet excellent treatment of the Unified and AASHTO systems (and some others) refer to Ritter and Paquette (1960), Chapter 6. For a complete description of the Burmister system, see Burmister (1953, 1955, 1958).

With definitions, discussion, and commentary regarding the importance of careful soil classification, the following list briefly gives some representative examples of valuable information pertaining to Case 1.

1. The potential borrow will be expensive to excavate because of its density and texture. Bulldozers or power shovels would probably be required. Scrapers would require "assistance" in excavating the soil.
2. Boulders (if confirmed) could create significant additional cost, depending upon frequency and size; better check road cuts.
3. The soil does not drain well (D_{10} size* in clay range). Hence, rain of significant amounts on borrow area would probably create delays and work stoppages, also possible trafficking problems. Check topographic maps for surface drainage patterns and access roads.
4. The soil has excellent texture from the standpoint of potential bearing capacity, depending of course, on degree (energy) of compaction, but will have poor drainage characteristics.
5. There is slight-to-medium potential frost action.
6. If compacted sufficiently, postconstruction settlements are no problem, but watch out for overcompaction, which could produce expansive potential because of highly plastic fines.
7. Best compactor is probably a rubber-tired roller.

1.4.2 Case 2

Case 2 involves a sand with coarser sizes predominant, fine gravel present, and no significant silty materials evident. (Sieve test showed fines fraction to be 7%; dilatancy test on fines confirms no plasticity NP.) Blow count range, 15–25.

*D_{10} size, also called Hazen's effective size, can be used to estimate soil permeability.

DEFINITIONS AND DISCUSSION. On the basis of preliminary inspection and evaluation, this soil appeared to "qualify" as a *select fill.* In New Jersey, one of the principal textural qualifications for a select fill is that it contain not more than 12% passing the No. 200 sieve. The sieve test was ordered to determine (particularly) the percent fines. The 7% confirms the select fill classification.

Geologically, this select fill is probably of alluvial origin. Judging by its low blow counts, it is a recent alluvium (mapped as AR on soil maps published by SCS, the Soil Conservation Service, a division of the U. S. Department of Agriculture) and has the following characteristics.

1. Excellent texture. It is becoming scarcer (except, of course, in shoreline areas) and is thus a valuable "commodity"; it should therefore be reserved for special purposes and used sparingly. Typical special purposes are trench bottoms to support pipelines and graded filters in earth-dam construction.
2. Free-draining. Thus, not susceptible to seepage pressures, frost action. If properly compacted, it is permanently stable (except Richter 8's).
3. Easily excavated (low blow counts, no cohesion).
4. No laboratory compaction test is necessary (or even sensible). Write compaction specifications on basis of percent relative density. Compact by flooding.
5. Best compactor would be a heavy vibratory roller.*

1.5 GLOSSARY

Alluvial A geologic term referring to soils transported or deposited by water (streams, glacial meltwater, etc.).

Dilatancy test A simple, diagnostic test of a soil to determine degree of plasticity. The finer fraction of the soil in question, generally that which passes the No. 40 sieve, is mixed with enough water to mold it into a soft, saturated ball about ½ in. in diameter. The ball is formed into a wafer shape and placed in the palm of the hand. The heel of the hand is then gently jarred repeatedly with the other hand, while one observes the surface changes (if any) of the wafer-shaped soil. A ready and quick shiny, glassy change in appearance (i.e., within seconds) constitutes a positive dilatancy test, and means that the soil is nonplastic (NP). If one squeezes the hand with the pat, the surface will immediately become dull in appearance; resumed tapping will quickly recreate the shiny appearance. (The squeezing shears the soil and allows the surface fluid to quickly drain away, thus assuming the dull appearance. The tapping consolidates the soil, and the water film reappears on the surface to yield the shiny appearance.) The quickly positive result

*I will explain these comments more fully in subsequent chapters.

identifies the soil as a "clean" silt (as opposed to, say, a clayey silt) or a fine sand/silt mixture. The longer it takes to produce any reaction (called dilatancy), the higher is the degree of plasticity. This and other field tests are used by experienced soils people to assess the plasticity rating of the soil in a qualitative way. For confirmation, the PI is determined by laboratory analysis for the liquid limit and the plastic limit, both of which are moisture contents. The numerical difference is the PI. (For details concerning field testing or laboratory testing, see Burmister (1953, 1955, 1958) and Lambe (1951), respectively.)

D_{10} size This means that 10% of the soil (by weight) is finer than that size. Thus, it may be thought of as an indicator of the fineness of a soil; an important index in assessing the general seepage quality of a soil. (See also Section 5.1.2.)

Frost action (frost heave) The seasonal heaving (winter) and recession (spring breakup) of frost susceptible soils, notably silty soils. A serious problem in highway and other route design. (See also Section 6.2.1.)

Graded filter A soil filter that provides a gradual (graded) transition for water flow, to prevent piping, for example. (See also Section 6.1.)

Granular (cohesionless) Coarser particles of soil (gravel, sand, "clean" silts) that exhibit no plasticity or cohesiveness.

Piping The dislodging and movement of finer soil particles under the influence of seepage of water through the soil, starting usually as a slow process, and leading eventually to greater distress and, possibly, failure (explained more fully in Section 6.2.1).

Plasticity (plasticity index, cohesive) Property of material (soil) that enables it to be deformed rapidly without volume change, without rupture, and without elastic rebound. The shape, of course, does change; a sculptor's modeling clay is a familiar example. The specific range of moisture contents over which the soil acts in this way is called the plasticity index (PI). All soils exhibiting a measureable PI are cohesive, meaning simply that the particles "stick together" naturally by other than (or in addition to) capillary forces. Clays, clay mixtures (clay with other soil types), and organic soils are cohesive.

Richter 8 The Richter scale is used to rate the intensity of earthquakes. Data obtained from seismographs located throughout the world results in the rating, which is on a logarithmic scale (base 10); thus a Richter 6 is 10 times more intense than a Richter 5, and so forth. A Richter 8, used somewhat facetiously in the text, would be a rare and very major earthquake of devastating effect.

Unconfined compressive strength The compressive stress at failure on a cylindrical specimen, usually expressed in tons per square foot (tsf); commonly performed on sampled, natural soils of a cohesive nature (e.g., clays). There is no reason, however, that such a test cannot be performed on compacted samples of suitable cylindrical dimension.

CHAPTER 2

Avoiding Costly Blunders

Before describing details of soil compaction and fill control, it will be helpful to focus briefly on a few broad topics, knowledge of which will help to avoid major, potentially costly mistakes.

2.1 THE PRACTICAL VALUE OF KNOWLEDGE OF THE HISTORICAL DEVELOPMENT OF SOIL COMPACTION

Ordinarily, expositions on historical development of a subject are presented in textbooks rather than practical guides. However, as I hope I will be able to demonstrate, the blunders that occur today in earthwork construction are strongly rooted in the past. While I am certain that Santayana was referring to much more esoteric subjects than soil compaction and fill control, his admonition that "those who cannot remember the past are condemned to repeat it" applies well.

Key periods include the 1920s and the early 1950s, which may be regarded as significant turning points in the development of the state of the art.

2.2 EARLY EMPIRICAL APPROACHES

Prior to the 1920s, indeed into antiquity, earthwork construction was strictly empirical. It is a matter of record, for example, that the Scot, John L. McAdam, stabilized soils by driving herds of sheep over areas of soft subgrades prior to construction of embankment and pavement sections (circa 1800). The modern sheepsfoot roller derives its name from this early construction practice. One can cite the Appian Way as a classic example of ancient road building of remarkable quality and stability, clearly accomplished without benefit of modern soil mechanics theory and practice.

Soil mechanics was started by Karl Terzaghi in the 1920s, with the decision of this great engineer–scientist to develop a scientific approach to treating soil as an engineering material. A contemporary of Terzaghi's, R. R. Proctor, is regarded by most geotechnical engineers as the originator of principles dealing with a rational approach to soil compaction, with his publication "Fundamental Principles of Soil Compaction" in *Engineering News Record*, August–September 1933.

2.3 RATIONAL APPROACH

Among Proctor's major contributions was a recognition of the need for a laboratory compaction test to control filling operations. Observing that soil moisture and compaction energy were two of the most important variables affecting compaction, he placed emphasis upon a determination of these relationships. Proctor's first decision was most probably the choice of a laboratory compaction energy that would simulate field energies available at the time, that is, 1920-vintage rollers. The method of applying the energy in the laboratory was through the use of a falling hammer, most likely chosen for computational simplicity rather than simulation of the kneading action of field rollers. (As is frequently the case, the exact simulation of field conditions in the laboratory is not practical or feasible, but it is essential to good testing practice to strive constantly toward this end, and to use good judgment in recognizing divergencies and their possible effects on results.) To simulate the field practice of compacting soils in lifts, Proctor specified that the laboratory test would simulate this factor by compacting in layers; thus laboratory layers simulate field lifts. Representative soil samples would be compacted in layers in a mold, at varying moisture contents, at energy chosen to simulate that of the field rollers.

2.4 STANDARD PROCTOR DENSITY

The most pertinent details of Proctor's original compaction test are shown in Table 2.1. By conducting this test at varying moisture contents, the laboratory curve for most soils (exception: free-draining soils) took the form shown in Figure 1.1. The form of the curve suggested the terminology—optimum moisture content—meaning moisture content that would produce 100% Proctor density or, more generally, moisture content that would result in the most efficient compaction in the field. Thus, if the natural field moisture happened to coincide with this value—not a likely event—field compaction efficiency would be ideal.

TABLE 2.1 The Proctor Compaction Test, 1920s

Mold Size	Layers	Hammer	Hammer Fall	Blows/Layer	Total Energy E_C
1/30 ft^{3a}	3	5.5 lb	12 in.	25	12,400 ft-lb/ft^3

[a] (4.6 in. × 4.0 in. diameter)

The engineers of the 1920s, undoubtedly recognizing the impracticality of specifying field compaction at precisely optimum moisture content, adopted a range of acceptable density values, as shown in the figure. Someone decided, apparently on a reasonable but totally arbitrary basis, that 95% Proctor would be acceptable. Thus the phrase "shall be compacted to 95% Proctor" was born. The contractor would have a reasonable range of densities (and field moistures) within which he could perform satisfactorily and efficiently. In practice, soils that were too wet could be dried by scarification, and soils that were too dry could be wetted by sprinkling, whereupon efficient compaction could be accomplished.

Field density tests of the rolled fill could be done to determine compliance with specifications.

2.5 MODIFIED PROCTOR DENSITY

In the late 1940s and 1950s, work progressed on the greatest road-building project in history: the 40,000-mile Interstate Highway System. By that time, it was possible to compact soils in the field to much higher densities needed to support the heavier vehicles that would use the system. This capability was a direct result of the development of a high-energy compaction test by the Army Corps of Engineers during World War II in connection with the need for heavy compaction for military airfields. According to Ralph Peck, the "principal players" in this development were Dad Middlebrooks, Jim Porter, and Arthur Casagrande. I am indebted to Dr. Peck for this information, and I am especially delighted to report it, inasmuch as I worked for Mr. Porter in my very first geotechnical employment in the summer of 1959. (Porter, Urqhardt, and O'Brien, Newark, NJ. Small world, isn't it!) And so, the subsequent use of the modified Proctor test for airport and road building was truly an example of "swords into plowshares." The "Goony Bird" became a DC-3, and the tanks became sheepsfoot rollers! Details are shown in Table 2.2. Note that the "new" test was simply a modification of Proctor's original test, hence the two names for the same test: modified AASHTO and modified Proctor. To further differentiate, the original Proctor was renamed "standard Proctor," and various organizations provided number designations. These are summarized in Table 2.3.

TABLE 2.2 The Modified Proctor (AASHTO) Compaction Test, 1950s

Mold Size	Layers	Hammer	Hammer Fall	Blows/Layer	Total Energy E_C
1/30 ft^3	5	10 lb	18 in.	25	56,000 ft-lb/ft^3

TABLE 2.3 Compaction Test Names

Names	Energy (ft-lb/ft^3)
Standard Proctor (or standard AASHTO)	12,400
ASTM D698	
AASHTO T99	
British Standard 1377: 1948	
Modified Proctor (or modified AASHTO)	56,000
ASTM D1157	
Modified AASHTO[a]	

[a] AASHTO is the modern name for AASHO, reflecting greater emphasis on transportation and transportation planning, notably mass transit in urban areas, upon essential completion of construction of the Interstate Highway System.

The names and designations for the tests should be used with precision, since the modified Proctor energy is about 4½ times that of standard Proctor.

2.6 LOAD-BEARING FILLS/BUILDING CODES

To further illustrate and emphasize the major difference in compaction energy of the two tests, you should be aware that standard Proctor densities have been historically regarded as *non-load-bearing*, at least in the sense of supporting structural loads such as footings. Building codes written before about 1947, for example, typically required all footings to bear on n.g.—natural ground. Standard Proctor densities represented the practical limits of construction capability at the time. Soils compacted to standard Proctor densities have some supporting capability, such as for parking areas and lightly traveled secondary roads, but they were never intended for support of heavier structural loads.

Today, it is common practice to specify footings to bear on compacted fill, typically compacted to 95% modified Proctor densities. According to Sowers (1979), modified Proctor energy is "comparable to that obtained with the heaviest rollers under favorable working conditions." I have performed field density tests on soils compacted to modified Proctor energies. I call them "blister densities," meaning that a person not used to such activity will develop blisters in the course of digging several holes with a garden spade. While this is not a very scientific terminology, the writer has found that this designation sticks most firmly in the mind of a young engineer assigned to the task.

2.7 SUMMARY

To avoid blunders, remember:

1. The major distinctions between *standard* and *modified*:
   (a) Standard energy: 12,400 ft-lb/ft^3.
   Modified energy: 56,000 ft-lb/ft^3.
   (b) Standard is *non-load-bearing* (light loads acceptable).
   Modified is *load-bearing*.
2. Use or interpret compaction test names (or number) designations carefully and precisely. Get angry (as I do) when someone says, "95% compaction."

2.8 GLOSSARY

Goony Bird This is the affectionate name given to the famous military cargo plane, the C-47, which "flew the Hump" in India during the Second World War, and later was the workhorse for the Berlin Airlift. Although one of these is on display at the Smithsonian Air and Space Museum, remarkably, *some of them are still flying!*

CHAPTER 3

Basics of Soil Compaction Curves: Laboratory Procedures

As a result of Proctor's work and subsequent developments in the state of the art by others, it is necessary to have a thorough knowledge of soil compaction theory and laboratory procedures in order to plan and execute construction of compacted fills in the field. Before proceeding to the field, then, we must look to the books and laboratories.

3.1 COMPACTION DEFINED

When a typical borrow is compacted in the laboratory, an interesting phenomenon is observed at very low moisture contents (0–5% approximately): the curve dips (see Figure 3.1). The moisture at this low level forms thin capillary films that develop tensile stresses upon impact of the compaction hammer. The tensile stresses cause a corresponding intergranular compression, resulting in a sudden increase in the effective strength of the soil. Figure 3.2 illustrates the situation schematically. As moldsful of representative borrow are compacted at higher moisture levels (2, 3, 4), the capillary films thicken and become an effective lubricant. At even higher moisture levels, however, a point of diminishing returns is reached, where the addition of more water results in decreasing densities (5, 6, 7). This behavior suggests what I believe to be the most illuminating definition of effective soil compaction: *the expulsion of air from the soil mass.* This definition allows for the easy understanding of an observation that is vital to good qualitative fill control in the field: *Soil undulation in front of and behind a field roller means the soil is too wet for efficient compaction at the energy level (roller weight) being used.* Figure 3.3 shows the analogy between the corresponding laboratory and field behavior. When the soil is being compacted efficiently (lubrication or "dry" side, Figure 3.1), air is ex-

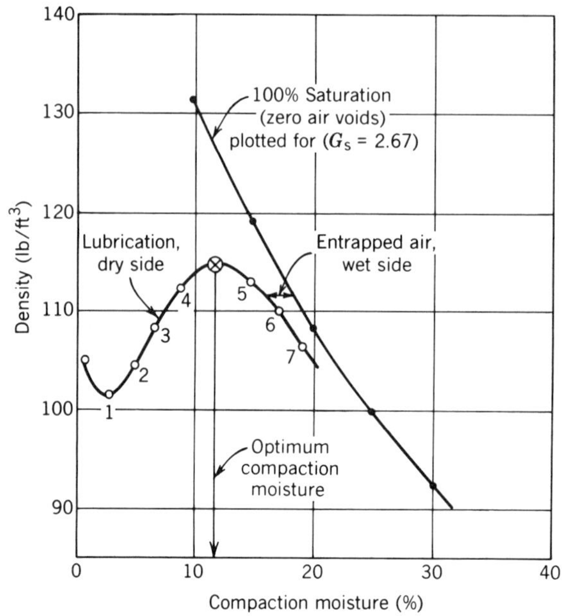

Figure 3.1. *Effects of moisture on soil compaction.*

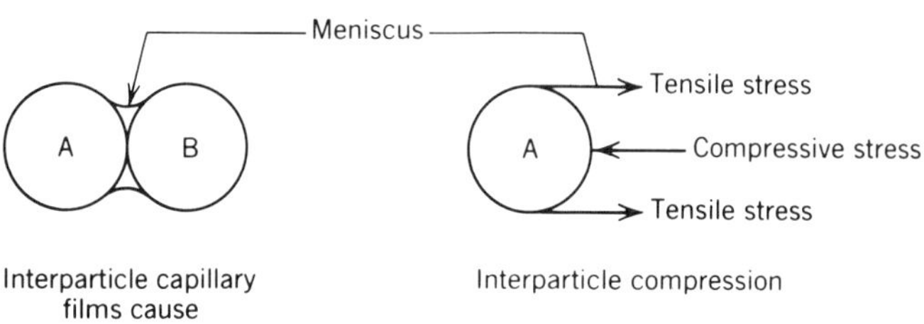

Figure 3.2. *Effects of capillary tension.*

pelled from the soil upon impact of the hammer *in quantities larger than the volume of water added.* However, soil has an "air permeability" such that at some high level of moisture (less air voids), the moisture cannot escape under impact of the hammer. Instead, the entrapped air is energized and lifts the soil in the region around the hammer. There are characteristic sounds associated with the laboratory compaction, a solid "thud—whoosh" (escaping air) for efficient compaction, and a not-so-solid "galumph" when the soil is too wet for compaction. In the field, the roller energizes the entrapped air in the soil, causing the undulation corresponding to the lifting of the soil upon hammer impact in the laboratory test. Efficient compaction in the field is, of course, characterized by an absence of undulation. The most helpful way of viewing what has happened when a soil is too wet is to recognize that water has been added without any expulsion of air, resulting in an increase in void space (i.e., a decrease in density).

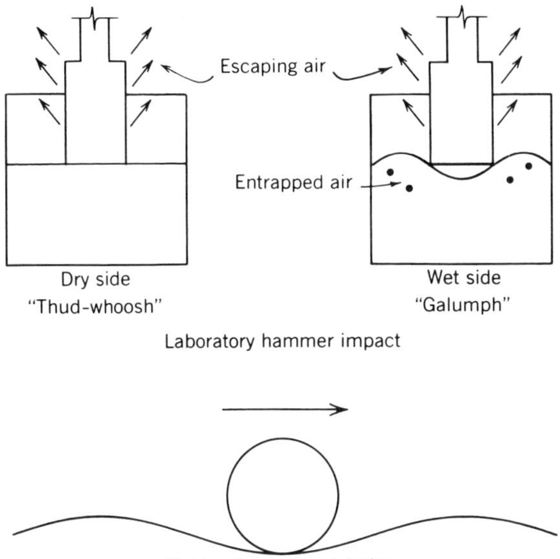

Figure 3.3. *Effects of excess moisture: laboratory and field correlation.*

3.2 SPECTRUM OF SOIL TYPES

Figure 3.1 shows entrapped air as the distance between the wet side of the curve and the line of 100% saturation (or zero air voids). The expression for dry density as a function of percent saturation S and compaction moisture content w_c is

$$\gamma_d = \frac{G_s \gamma_w}{1 + \dfrac{w_c G_s}{S}}$$

where γ_d = the dry density of the compacted soil
G_s = the specific gravity of soil minerals
γ_w = the unit weight of water
w_c = compaction moisture content
S = percent saturation

For cohesionless soils (quartzitic), $G_s = 2.65$, and for most common clay minerals, $G_s = 2.70$. With $S = 1$, a zero air voids curve can be plotted which, with little error, would be representative of all soils. Furthermore, the zero air voids curve is an envelope under which the compaction curves for all soils, compacted at any energy level, will fall, with wet sides approximately parallel to the saturation line envelope. An array of such curves is shown in Figure 3.4. Curves for five soil types, intended to represent the spectrum of all soils commonly used in earthwork, are shown, each compacted to the two common energies: standard Proctor and modified Proctor. Curves 1 and 5 represent TALB: typical allowable load-bearing borrow. Texturally, it would ordinarily be well-graded and predominantly granular, but with significant percentage of fines, preferably of sufficient plasticity to impart overall cohesive qualities (i.e., "clay binder"). In my suggested auxiliary classification system described in Chapter 2, it would be an EGS, and would be marginally draining or impervious, depending principally on the exact quantity and plasticity of fines. Note that such a soil can be compacted to densities approaching 140 pcf with the modified Proctor energy: "blister densities."

Curves 2 and 3 depict free-draining sands, or select fills. Curves 6 and 9, and 7 and 10, represent "lean" clays and "fat" clays, respectively. "Lean" and "fat" are commonly used by soils engineers to designate relative degrees of plasticity. Thus, a fat clay is highly plastic and, for our purposes here, we can say it has a greater affinity for water. I call clay mineral particles of this type "thirsty little devils." Lean clays have low to medium plasticity, with PI's less than 40 and liquid limits less than 50%. Casagrande's Plasticity Chart is the most common method of classifying plastic soils (Lambe, 1951, p. 27).

Finally, curves 4 and 8 represent silts, fines that are nonplastic (NP) or, at most, slightly plastic (PI 0–5).

In summary, as strongly inferred by the preceding definitions, it is important to know not only the amount (percentage) of fines in a soil mixture, but also the plasticity of those fines. Construction problems and costs can be markedly different for different textures. The engineering behavior of the compacted soil also varies greatly for different textures.

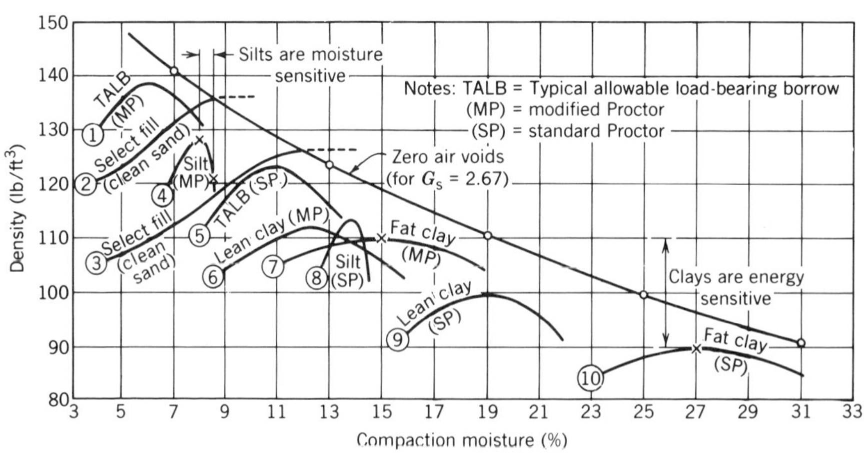

Figure 3.4. *Compaction curves for spectrum of soil types.*

3.3 CURVE LOCATIONS AND SHAPES AND THEIR PRACTICAL MEANING: MOISTURE AND ENERGY EFFECTS

Of all the illustrations in this book, I regard Figure 3.4 as the most important; important enough to be studied over and over until its details are fixed in memory, including approximate numerical values. Following are explanations pertaining to details.

3.3.1 Variations in Optimum Moisture Contents

The optimum moisture content of a soil is inversely related to compactive energy. (There is a widely held misconception that a soil has an "inherent," unchanging optimum moisture content. This misconception probably developed because, for the early years (1920–1950), only one energy was used, standard Proctor.) *A soil can have almost any optimum moisture content, depending upon the compactive energy used.*

Although the range of compaction moistures shown in Figure 3.4 extends from 3 to 33%, observe that the range of optimum moisture contents for TALB has a considerably narrower range, about 5–12%. Moreover, if the borrow is to be load-bearing (as defined earlier), it should be compacted to energies approximating modified Proctor, narrowing the range to, say, 5–9%. Thus, with knowledge of the foregoing, one could estimate the approximate optimum rather closely. Indeed, if one commits the set of curves shown to memory, even approximately, *a fairly close estimate could be made for the optimum moisture content for any soil compacted at any common energy level* (i.e., standard or modified Proctor).

In the foregoing commentary, note that the term TALB refers initially to the textural quality of the fill. The other criterion for load-bearing capability is, of course, the *condition* of the fill—its compacted density. The literature frequently refers to controlled fills and uncontrolled fills without further distinction, but it is helpful and frequently of considerable practical (cost) consideration to be aware that a fill can be controlled in two ways: quality (texture) and condition (placement density). This matter is discussed more fully in subsequent chapters.

3.3.2 Moisture and Energy Sensitivity

The shapes of curves have practical significance. Silts and clays are two extremes, with silts steep and clays relatively flat.* As indicated on the figure, *silts are moisture-sensitive*, meaning simply that a small change in field moisture will effect a major change in compacted dry densities achievable *at the same*

*Throughout this broad discussion of the entire spectrum of soil types, the terms "silt" and "clay" refer to soils that are essentially silt (or silty) or clay (or clayey), as defined earlier. A more specific treatment of percentages of various soil constituents and their effects is presented in Chapter 5.

compaction energy. (Note that two field options for correction would be possible: changing the soil moisture or changing the compaction energy.) *Clays are energy-sensitive*, meaning that relatively small changes in compaction energy can produce large changes in density (curves 7 and 10). Thus, if one area of a site is inadvertently overcompacted by channelization of construction traffic, larger future differential settlements will result. Stated another way, it is probably better to have the entire fill compacted to, say, 92% modified Proctor than half 92 and half 98. In fact, serious overcompaction of clays can produce a fill that will expand on future wetting, thus subjecting any structure on the site (or the fill itself, as with earth dams) to intolerable deformations.

3.3.3 Free-Draining Soils: Select Fills

Free-draining sands, or select fills, have no discrete optimum moisture content (curves 2 and 3), for the simple reasons that they *are* free-draining. Thus, if one attempts to generate a compaction curve of the more typical peaked, symmetrical shape, the added moisture, at higher levels, would simply drain or be ejected from the mold upon impact of the hammer. In the field, therefore, fills must be compacted either completely dry or by flooding. Except in very arid areas, the former condition is impractical, so flooding is normally used, a process of compacting the soil by applying an excess of water, usually with a hose.

Comprehensive details regarding field compaction of all soil types will be presented in Chapter 6, Fills and Fill Compaction.

3.4 ASTM COMPACTION REQUIREMENTS

To conclude this section, preparatory to a detailed presentation of the nature of the problems in earthwork construction, it will be helpful to describe briefly the steps generally required to develop laboratory compaction curves needed for field compaction control. These steps generally follow those required by ASTM, the American Society for Testing and Materials.

1. Obtain a large bag sample of the soil at the site or borrow area that is judged to be representative. Generally, 100 lb or more is required to prepare several batches at different moisture contents. Recompaction of the same batch, by extruding a compacted soil, pulverizing it, and adjusting to a different moisture level, is not allowed. Research has shown that significant errors can occur if the recompaction process is used. The error is probably more pronounced at higher compaction energy (modified Proctor), since the higher energy would undoubtedly cause more "chipping" of larger soil particles upon repetitive recompaction. Angular soils are probably more susceptible to such chipping. Modification of the soil in this way during recompaction in effect produces a different soil (grain size distribution) upon each recompaction. Since the soil in the field will not be so modified, the notion of simulating field conditions is violated without justification. The only ap-

parent justification is the ease of handling and time saved recompacting a small (10 lb) representative sample.

Successive recompaction of cohesive soils should also be avoided because an unrepresentative rearrangement of soil particles would result, creating a more pronounced parallel layering than would be the case for soil compacted in the field. This layering is called a "dispersed" structure. Its effects are described more fully in Section 6.2.2.

2. Air-dry the soil overnight, spreading the soil out to a depth of 3–4 in. to facilitate drying.

3. Separate the soil into approximately equal, representative batches, by a quartering or splitting process.

4. Determine the weight and moisture content of each batch.

5. Determine the dry weight of soil retained on the No. 4 sieve, and record as percentage of batch. The maximum particle size used in the compaction mold is logically based upon mold size. If a very large percentage of "No. 4" is present, it is sensible to use a larger mold and a correspondingly larger exclusion size. A 6-in.-diameter mold, with a +¾-in. exclusion, is common. If this option is exercised, the energy should be increased to about 55 blows per layer for the modified Proctor test. To account for the material excluded in the compaction test, a correction should be made to field density values, based upon differences in excluded material between the laboratory and field (density hole) samples.

6. Add moisture to each air-dried batch to produce moisture contents in 2–3% increments in the anticipated range encompassing optimum moisture content. Considerable experience is necessary to accomplish the foregoing in an efficient manner, ideally producing two points on either side of optimum. However, consideration of Figure 3.4 and related explanations should enable one to estimate the approximate moisture contents required. To effect the actual soil preparation at the desired moisture contents, an instant moisture balance, used with the following formula, is very helpful:

$$W_A = \frac{w_d - w_k}{1 + w_k} W_{TB}$$

where W_A = weight of water that must be added to a soil of known moisture content to raise it to some higher (desired) moisture content

w_d = desired moisture content

w_k = known moisture content (as determined by the "instant" moisture balance*)

W_{TB} = total weight of wet batch of soil (i.e., soil + water)

*Cenco Model No. 41101.

The instant moisture balance is a table-top device that enables one to determine the moisture content of a soil in several minutes. In the typical case, one would determine the air-dried moisture content, w_k (known moisture content), and then plug into the formula successively the values of moisture content desired w_d to determine W_A values necessary to generate the curve. Because of the need for thorough mixing of each batch, which will diminish the moisture content by aeration, it is necessary to overshoot the desired moisture content. Two to 3% overshoot is recommended. Thus, if one wishes to produce a thoroughly mixed soil of moisture content $w_d = 9\%$, use $w_d = 11.5\%$ in computing the amount of water to be added. These procedures are also very helpful for filling any "holes" in the compaction curve that are discovered after plotting the curve. Alternative procedures may introduce unnecessary errors, or are time-consuming and frustrating (trial-and-error, the usual method).

7. Place the wetted, prepared batches in a large can with a lid or in sealed plastic bags. Cure overnight in a humid room. (This allows the moisture to soak into the soil particles, thus providing closer simulation of field conditions.)

8. Compact each batch in accordance with the required energy (e.g., Table 2.1 or Table 2.2). Determine the weight of each compacted soil sample and, from the extruded cylinder of soil, obtain representative chunks for determination of compaction moisture w_c. Weigh chunks and place in oven (105°C) overnight, or until dry weight does not change.

Compute the dry density:

$$\gamma_d = \frac{W/V}{1 + w_c}$$

where W = weight of soil in mold

V = volume of mold

w_c = compaction moisture (expressed as decimal)

9. Plot the compaction curve and zero-air-voids curve (see Figure 3.1).

3.5 SUMMARY

1. Undulating fill means that the fill is too wet for the energy of the compactor being used.

2. Optimum moisture content of a soil decreases with increasing compaction energy.

3. Figure 3.4 is an excellent aid in estimating the approximate range of optimum moisture content for any soil.

4. Silts are moisture-sensitive (steep curve).

5. Clays are energy-sensitive.

6. Select fills have no discrete optimum moisture content.

7. ASTM laboratory compaction requirements are extensive and time-consuming. Counting acquisition of the sample and delivery time, one day each for air-drying, curing wetted samples, and moisture content determinations, a minimum of four days would be required. With other common delays, such as those caused by heavy laboratory testing demands, a full work week is often required to obtain the results of one test. This feature of compaction testing can become a matter of considerable importance, and relates to one of the major problems in compacted fill technology, that of changing borrow.

3.6 GLOSSARY

Compaction moisture content, w_c The moisture content w of a soil is the ratio of the weight of water in the soil to the weight of the solids (minerals), expressed in calculations as a decimal, but often referred to otherwise as n%. The compaction moisture content, then, is simply the moisture content at which the soil is compacted.

Percent saturation, S The ratio of the volume of water in a soil to the volume of voids, expressed as a decimal in calculations.

Quartering A technique used for reducing a large sample of soil to a representative sample of a desired smaller size, done by spreading the soil to about a 3-in. deep pile, and cutting it into quarters, then one quarter into quarters, etc.

Splitting Another technique for obtaining representative samples of desired size. In this case a "soil splitter," an apparatus designed for the purpose (available from various suppliers of laboratory equipment), is used.

CHAPTER 4

Major Problems in Compacted Fill Technology: Proposed Solutions

Fill control is the art of ensuring the proper selection, placement, and compaction of soil to provide permanent stability for the structure it is to support. This control is usually provided by preparing written specifications governing the texture and compacted density of the fill. End-result specifications normally call for some percentage of the laboratory-determined maximum of some recognized standard, for example, 95% standard Proctor or modified Proctor. Methods specifications deal with matters such as required field compaction moisture, lift thicknesses, and compaction equipment. Almost all specifications incorporate features of both types requiring a *target value density* and specifying means of attaining that value. Of logical necessity, however, some flexibility should be incorporated in methods specifications to accommodate day to day changes in the field (notably weather, and thus moisture); methods of attaining the target value will require adjustment.

Following are descriptions of problems in fill control, their causes, and some suggested remedies. Some brief case histories are included to emphasize some of the major problems, problems that all too frequently involve major blunders resulting in increased construction costs, construction failures, or unpleasant confrontations leading to professional embarrassment and, in one case, unethical behavior.

4.1 STANDARD-MODIFIED IGNORANCE

Because of the relatively recent transition from standard Proctor to modified Proctor energy, as described earlier, there exist many old specifications in the job files of many engineering firms calling for Proctor densities (now called standard Proctor). These, of course, were written for earth dams, levees, roadway embankments or parking areas, or for some purpose other than support of structural foundations, since building codes did not generally permit such practice. Also, because of the very newness of soil mechanics itself, there are many older engineers who did not take formal courses during their college years and, for various reasons—mostly an understandable preoccupation with their own specialty interests—have not become aware of the very significant difference between the two terms, standard and modified.

Of such a combination of circumstances are major blunders made.

4.1.1 Case Study 1

In the early 1960s, I was assigned to a fill control job involving the placement of about 150,000 yards of fill.* The assignment was made on something of an emergency basis. The soils consulting firm for which I worked received a phone call on Friday, requesting the service of a fill-control specialist for the following Monday morning. I was given the address of the firm and the name of the engineer requesting the service. No other information about the job was provided.

Within a few days of arrival, the following situation and information was discovered:

1. Stripping and grubbing had been completed (removal of trees and brush).
2. Site preparation involved a cut—fill operation to convert a gently sloping site into a flat area, with the fill constituting approximately 60% of the site.
3. Filling operations had already started. One compactor was available, a small steel-wheeled roller of obvious early vintage, including old-fashioned spoked wheels and a gravity-feed gas tank.
4. No borings were available.
5. No investigation had been made to locate borrow. In fact, it appeared that the problem had not been considered.
6. Written specifications, amounting to about two-thirds of a typed page, were supplied. Ninety-five percent standard Proctor density was the specified target value for the fill.
7. The thickness of the fill varied from zero (at the no-cut, no-fill line) to approximately 14 ft at the downhill region of the site.

As you may infer, the situation described was to be fraught with difficulties. However, in order not to lose focus on the purpose of this section—major problems of fill control—descriptions of routine problems and operations are deferred to subsequent chapters.

*A common slang term for cubic yards.

About two weeks into the job, an engineer (not a soils specialist) representing the ultimate site tenant visited the site, principally to observe site preparation activities. During an early conversation, a comment was made about 3000-lb/ft^2 floor loads.

It became evident that something was seriously wrong: 3000-psf floor loads on a 14-ft fill were not even remotely compatible with 95% *standard* Proctor energy. The local consultant's engineer suggested that the necessary higher densities could be obtained by getting the contractor to use heavier rollers and more passes—that, if this were done cleverly, the contractor could be made to produce *modified* energy densities without being apprised of any specification changes.

Telephone consultation with my project engineer led quickly to an agreement to ignore the unethical proposal, and a job meeting was arranged for the purpose of addressing the problem and its ramifications. The specifications were rewritten with the major change to modified Proctor energies, and the contractor was properly compensated for the required extra work.

As might be expected, there were no conversations among the parties concerned regarding the unethical oral proposal described. Thus, the circumstances leading to this blunder must—with one exception—be surmised. It is certain that the engineer to whom the writer reported, and who supplied the "original" specifications, knew very little about soil compaction. The specifications were probably obtained or derived from the files. Finally, it is speculated that the retention of the firm to supply consulting services and inspection for soils work dealing with site preparations was a last-minute decision based on a late-developing awareness of the need for such services.

As may be inferred from these circumstances, the earthwork contractor's superintendent was also largely ignorant of the geotechnical engineering aspects of soil compaction and fill control. As surprising as this may seem, in view of its overriding importance to successful bidding, it has been my experience that such ignorance is common. Even big earth work contractors sometimes lack even a rudimentary knowledge of the major differences between standard and modified Proctor, notably differences in cost and effort. Many small contractors cannot achieve modified Proctor densities commonly called for in specifications for load-bearing fills, because they do not own, nor can they afford, the heavy rollers needed to produce such densities efficiently. (See Case Study 2, following.)

The solution to the problem of standard–modified ignorance is to disseminate knowledge widely among persons who are not soils specialists and who are involved in earthwork. This can be done effectively in two ways: (1) through publication of concise, practical information in a practice-oriented medium that is explicitly advertised as "useful in practice" and is widely marketed,* and (2) by convincing academicians of the importance of greater

*The phrase "useful in practice" is an officially recognized keyword in the Geodex Information Retrieval System. It is, however, unlikely that very many persons who are not soils specialists have access to this system or, indeed, are even aware of its existence.

emphasis on the subject in courses that are composed of a large majority of students specializing in areas other than soils (i.e., structures, construction, transportation, and "other"). This class composition would always be the case in the undergraduate *required* course in soil mechanics offered at accredited engineering colleges. Moreover, records compiled over a period of seven years in a graduate course, "Shallow Foundations," at the New Jersey Institute of Technology (NJIT) indicate that 77% were nonspecialists (86 of 111). Data of recent years show at even higher percentage: over three years, 84% were nonspecialists (62 of 74).

4.2 THE NINETY-FIVE PERCENT FIXATION

Since the reasonable but otherwise arbitrary choice of 95% Proctor densities as the target value for fill control work of the 1920s was made, there has developed a fixation on the part of specification writers to require this percentage compaction, irrespective of loadings, fill thickness, or other factors that should logically influence compaction requirements. There are a number of causes for this practice: a reluctance to specify anything different than is "customary," nonspecialists writing compaction specifications, and the fact that there is no widely accepted rational method for specifying percentage compaction appropriate for specific conditions.

4.2.1 Case Study 2

A small project involved the expansion of a structure for light industrial use. The proposed expansion was to extend to a sloped area as shown in Figure 4.1. I recommended orally that a continuous wall footing be founded in compacted fill (level A), thus avoiding the apparent greater cost of a couple of lifts of concrete block (for footing at level B). The contractor agreed.

I performed a geotechnical site investigation and prepared a report that was distributed to the structural engineer and the contractor. The pertinent excerpt from my report is: "It is recommended that this fill be placed in lifts 8 in. to 12 in. thick and compacted to *93% modified AASHO*."

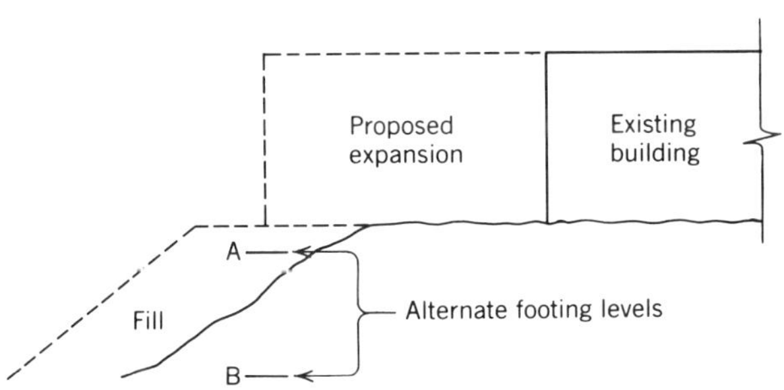

Figure 4.1. *Case Study 2.*

In time, after receiving my report, the structural engineer prepared the structural drawings, among which was included the foundation plan containing the following:

> *Foundation Notes*
> 1. The contractor shall read the soils report dated . . . by Edward J. Monahan, P.E. and shall familiarize himself with the site and the conditions as outlined in the report. . . .
>
> *Compacted Fill Notes* . . .
> 4. Fill shall be placed at its optimum moisture content in uniform layers not more than 8 in. thick after compaction, and each layer shall be thoroughly compacted to a density not less than *95 percent* of the density prescribed in ASTM D1557-66 T. [emphasis mine]
>
> 5. The compacted fill shall be placed under the supervision of a licensed professional engineer . . . who will submit a certification of the fill.

COMMENTARY Because of the light loadings and relatively thin fill required, I recommended less than the usual percent compaction, that is, 93 rather than 95. Note, however, that the structural engineer, while acknowledging the writer's geotechnical expertise (Foundation Note No. 1), could not accept the "unusual" lower compaction percentage recommended.

Also note that the specifications called for a "certification of the fill" by "a licensed professional engineer" and "placement supervision" (Compacted Fill Note No. 5). As it developed, the contractor ignored the supervision requirement. The fill was placed and compacted, a continuous footing was poured, and several lifts of block were constructed, all without the specified supervision. With some misgivings and dismay, and with certainty about the outcome, I performed a field density test on the compacted fill in the area immediately adjacent to the wall. As expected, the fill was nowhere near the specified 95% modified AASHO density. In fact, the field density test indicated approximately 67% modified AASHO. As a result, I had no recourse but to insist upon the costly procedure of ripping out the wall and the fill, and rebuilding the wall foundation in accordance with the specifications for placement and certification.

In addition to the principal illustration of the "95% fixation" problem, this simple case history presents evidence of the other problems: (1) Many contractors do not generally appreciate or understand what the stringent requirement of modified energies (at high percentage levels) entails in terms of required machinery and placement methods (moisture, lift, thickness, number of required passes); (2) nonspecialists prepare unrealistic soil specifications—unrealistic with respect to compatibility with load requirements and unrealistic with respect to demands on contractors. In addition to the unnecessarily high energy requirement, the phrase "*at its* optimum moisture content in uniform layers not more than 8 in. thick *after* compaction [emphasis mine]" contains questionable verbiage. What does "*at its* optimum moisture content" mean? Taken literally, *any* divergence from optimum is not permissible. Since the important end-result requirement is 95% modified AASHO density, some latitude in methods should be permitted. Indeed, it is logical that completely rigid methods specifications make it impossible to comply with an end-result specification!

The phrase *its* (optimum moisture content) makes one suspect that the writer of the specifications believes that the soil has *an* optimum moisture content—a typical misconception of nonspecialists. (See earlier commentary, Chapter 3, Basics of Soil Compaction Curves.)

The phrase "8 in. thick *after* compaction," again taken literally, would require the fill inspector to reject the fill if the compacted surface elevation revealed said "violation," even if the soil density target value was satisfactory. Lift thickness requirements should therefore specify an acceptable range of *loose* placement thickness.

Finally, one wonders what the word "thoroughly" means in Compacted Fill Note No. 4.

These criticisms may seem niggling compared to the more serious error of over- and underspecifying compaction energies, but collectively such careless phrasing in specifications can create unnecessary friction in field operations.

4.2.2 Case Study 3

A large construction project for a college medical complex included structures and parking areas with a wide variety of loading and site preparation requirements. The resident soils engineer, an evening graduate student in my course, "Shallow Foundations," raised a question in class that had developed on the job that day. Work had concluded in one area of the site that involved compacting soil to 95% modified Proctor density, and attention was shifted to another area of the site requiring fill. In this case, however, apparently because of lighter loadings and other factors allowing for less stringent fill compaction, the written specifications called for 95% *standard* Proctor density. The problem was that no compaction test had been done for other than modified Proctor energy, so no target value could be determined; that is, no standard Proctor curve was available. The question: What to do? The recourse of the student (resident engineer) was to call his office for advice, which was to use 93% modified. The engineer asked his supervisor, "Why 93%?" The answer: "Because I said so."

It is evident that 95% standard Proctor is equivalent to some lesser percentage of modified, but no clear-cut, rational way appeared to exist to determine the specific lower percentage for a given soil; in effect, an experienced person simply uses judgment. Thus, in Case Histories 2 and 3, the judgment was, in each case, to use 93%.

One means of correlating standard and modified percentages is simply to compile data on a variety of soils that have been compacted at both energy levels. Thirty-three soils were thus evaluated by undergraduate students at

NJIT (Mardekian, et al., 1973). The percentage modified corresponding to 95% standard ranged from a low value of 81.4 for a "clay" to 92.2 for a "Florida sand," with an average of 86.6. Only 4 of the 33 samples exceeded 90%, each of which was an essentially granular soil, predominantly sand. Accurate textural descriptions and related index properties were not obtained for most of the soils, but it appears safe to conclude that, in general, 95% standard Proctor conforms to percentages considerably less than the 93 value used in Case Histories 2 and 3. Since the level was chosen in both cases strictly on the basis of experience, the figure of 93 was "sensibly conservative."

As a result of my concern with the problems described, I developed an explicit method for specifying percentage soil compaction (Monahan, 1974). Because of research and additional analyses, certain improvements have been made. Following is a description of the method, with explanations of the need for further modifications, and suggestions for research to improve and extend the method toward broader applicability in practice.

4.2.3 A Method for Specifying Percentage Soil Compaction

Almost all site preparation work involves the compaction of soil. And many specifications for such work require 95% modified AASHTO densities, irrespective of the intended use of the fill. In cases such as lightly loaded warehouses, parking areas, and subgrades and embankments for secondary roads, such a stringent compaction requirement is unnecessary and consequently not economical. In special cases, where heavy loads and/or structures highly sensitive to differential settlements are involved, it may be advisable to specify a percentage compaction in excess of the "the usual" 95%. To provide a means for determining economical compaction requirements, the following procedures have been developed.

Figure 4.2 is the familiar design chart recommended by Terzaghi and Peck for footings on sand; allowable bearing pressures indicated are based upon the settlement criterion (which governs for all but the loosest sands and/or very small footings). The usual use of the chart is by abscissa-entry; that is, one enters the chart with a first-estimate footing size, and reads the allowable bearing capacity corresponding to the *in situ* blow count N for the natural soil deposit as determined by the Standard Penetration Test. (*Note:* Some corrections to the blow count may be necessary.) The essential difference between this procedure and one where compaction is to be performed is that control can be exercised over the eventual bearing capacity of a controlled fill. Thus, one can reverse the procedure and ask,

> What "blow count" is needed to provide a specified (or desired) bearing capacity, and what percentage compaction is needed to develop a density which would yield such a blow count?

Figure 4.3 suggests a method for answering such a question. The graph represents the results of a compaction test for a TALB, the most common of which

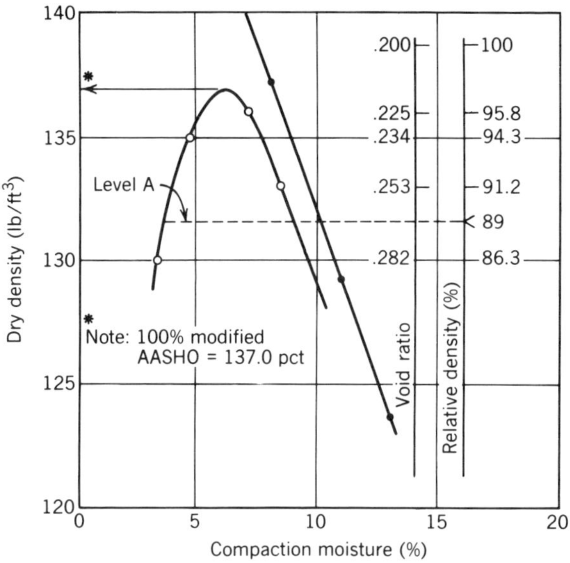

Figure 4.2. *Allowable bearing capacity for footings on sand. From Peck, Hanson, and Thornburn, Foundation Engineering, 1st ed., Copyright John Wiley & Sons, 1953.*

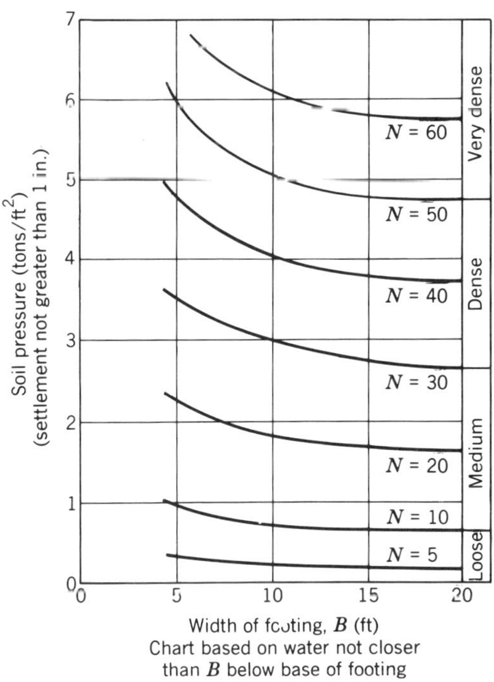

Figure 4.3. *Modified AASHTO compaction test for a TALB.*

is modified AASHTO (ASTM 1557). Added to the plot is a scale of void ratios, with limiting values of e_L and e_D corresponding to relative densities of 0 and 100%, respectively (loosest and densest). The relative density is given by

$$D_R = \frac{e_L - e_n}{e_L - e_D}$$

where e_n is usually thought of as the natural (*in situ*) void ratio. However, in this procedure, the value is interpreted as the required void ratio of the compacted fill e_c:

$$D_R = \frac{e_L - e_c}{e_L - e_D}$$

It remains to define procedures for estimating e_L and e_D. For very large, important jobs, involving either large cost or protection of human life (as with earth dams or high retaining walls) ASTM procedures for e_L and e_D determinations are recommended. However, as a practical matter, it is thought that quicker, approximate, and much less expensive methods are justified for most routine applications. Indeed, it has been my goal to develop a design method that will combine validity with acceptability to users. Other criteria are simplicity, quickness, and low cost. Finally, in the special but not too uncommon situation where the borrow texture is changing frequently—that is, from day to day on a job—it is impractical to use anything other than quick, simple, low-cost methods for developing criteria for fill control specifications. (This problem and proposed methods of solution are presented in more detail in a following section, Changing Borrow.)

ZERO PERCENT RELATIVE DENSITY The loosest possible practically obtainable density of a granular soil can be determined approximately by allowing an oven-dried sample of the soil to fill a container of known volume by gravity flow with little free-fall. This can be done by fashioning a miniature tremie by connecting a length of rigid plastic hose to a funnel. The oven-dried soil is then placed carefully, avoiding free-fall, into a 1000-mL plastic graduated cylindrical flask. The weight and volume of the soil gives the approximate loosest dry density, from which e_L may be calculated:

$$e_L = \frac{G_s \gamma_w V}{W_s} - 1$$

where G_s = 2.65 (for granular soils, closely)

γ_w = the unit weight of water, 62.4 lb/ft^3 or 1 g/cm^3

V = the measured volume

W_s = the measured weight

ONE HUNDRED PERCENT RELATIVE DENSITY The densest condition can be approximated by vibrating the same sample of oven-dried soil until no further significant reduction in volume is observed. In the laboratory, this may be done most easily by securing the base of the flask to a mechanical vibrator. (An inexpensive Sepor Vibra-Pad has been used successfully.) The plot of periodic readings will establish the approximate minimum volume, and e_D may be calculated as above.

If significant amounts of clay are present, it may be necessary to pulverize dried clay lumps with a mortar and pestle before consolidation by vibration. Where large amounts of clay are present, an alternate method of establishing e_D may be adopted, as shown by Figure 4.4, which illustrates schematically the general form that compaction data would take for a TALB compacted at different energies. The quantity $\Delta\gamma$ is the difference between the maximum dry density for modified Proctor and that corresponding to 100% relative density. Research to date suggests that this increment is a relatively small number, perhaps 2–3 pcf, for TALBs. (While no testing has been done, it is thought *not* to be true, in general, for other soil types, especially ECS or "pure" clays.) Thus, to establish e_D without serious error for TALBs, one could merely add 3 pcf to

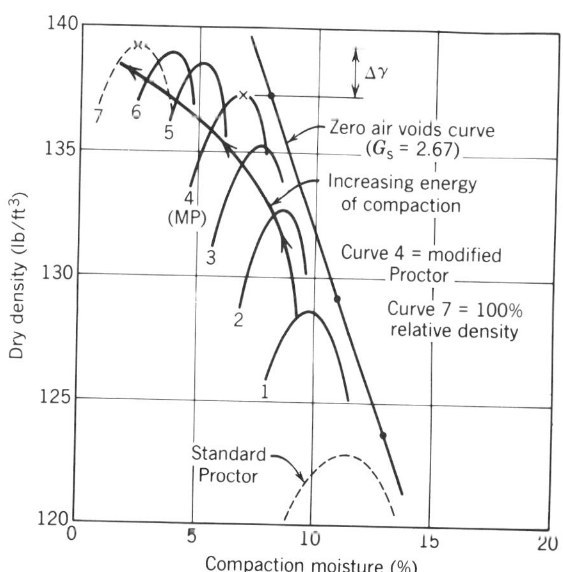

Figure 4.4. *Establishing* $\Delta\delta$.

the 100% modified Proctor value. When time permits, the experimental approach of establishing $\Delta\gamma$ is recommended. As more data accumulate for a wide range of textures, the validity of assuming $\Delta\gamma$ can be determined.

FIELD DETERMINATION Similar procedures may be followed in the field, as necessary. A camp stove may be used to dry the soil, and vibration may be effected manually by repeated gentle tapping of the plastic cylinder with a hammer. A small confining surcharge weight is recommended.

PERCENTAGE FIELD COMPACTION To determine percentage compaction relating to the fill control target value, relative densities must be related to blow counts. Table 4.1 lists relationships suggested by Burmister.

LIMITATIONS Figure 4.2 and Table 4.1 are applicable for foundations on sand. Extending the method to a TALB (Figure 4.3), as in the example, is therefore open to question. However, as all practicing engineers are aware, it is the essence of good engineering to exercise good judgment in the use of whatever methods are available. Accordingly, it is recommended that the method presented here is reasonably applicable to all soils that can be considered EGS, essentially granular soil, including those like many glacial tills that contain binder fractions. Where the clay is of sufficient quantity and plasticity such that the granular particles are judged to "float" in the clay matrix, the method should not be considered for use in design. The method, therefore, is definitely not suitable for clay fills. Another guiding criterion might be to consider whether intergranular contact between granular particles exist to a significant degree in the soil.

The question of applicability is often not an easy one to answer. However, it may finally be noted that any rational method, of even limited applicability, is better than no method at all.

EXAMPLE A fill is to support a 10-ft-wide footing and is to have an allowable bearing capacity of 4.5 tsf. What percentage modified AASHTO is required?

TABLE 4.1 Blow Count/Relative Density Relationship of Sands (Fairly Reliable)

	Relative Density	
Number of Blows per Foot N	Designation	Percent Range
0–4	Very loose	0–10
4–10	Loose	10–40
10–30	Medium	40–70
30–50	Dense	70–95
Over 50	Very dense	Over 95

Source. After Burmister (1955).

Look at Figure 4.2 and determine that a blow count of $N = 45$ is required. By interpolation, this corresponds to $D_R = 89\%$.

Assuming that Figure 4.3 represents the modified AASHTO compaction test on the borrow, and further that the void ratio and relative density scales have been determined as described, it is found that the required relative density corresponds to level A, which is 96.1% modified AASHTO (131.6 pcf).

Thus, the specifications for this fill could be written to require 96.1% modified AASHTO compaction. Clearly, it would then be a simple matter to determine target values for any other fills, using the specific requirements for fill support and footing size.

4.2.4 Case Study 4

A recent case study, ending in 1990, involved a job where housing developments were to be constructed over an area that had earlier been backfilled for a sewer construction project. Parts of the sewer job had been constructed through open farm country, and included some rather deep cuts. Thus, the areal extent of the backfill surface was considerable. The specifications for the sewer job sensibly called for 95% standard Proctor densities. Unfortunately, provisions were not made for continuous on-site inspection, and it was contended in extensive litigation that the sewer backfill was not properly compacted. Numerous experts were brought in over a period of years to evaluate the situaiton, involving both the examination of existing records (such as they were), and subsurface postconstruction site investigations of considerable sophistication. These investigations were done largely as a result of the paucity of accurate construction and inspection records from the original sewer work.

The cost of this extended litigation, as measured in attorneys fees, experts fees, and site investigations, was "in the high six figures," as told to me in some wonderment by the attorney who retained me. (Guess who got most of the money!) He clearly believed that the escalating costs had reached absurdly high levels.

One of the puzzling, even bizarre aspects of the situation was that the specifications for the residential development were calling for 95% modified Proctor densities and even higher, apparently in the belief that such extraordinarily high compaction requirements would somehow make up for the suspected inadequate compaction of the underlying sewer backfill. I never did understand this, and said so repeatedly in depositions. It was sort of like specifying a special high strength titanium alloy for the fuselage of an airplane to make up for the suspicion that the engines wouldn't work.

The case was finally settled before trial.

Following are some of the geotechnical aspects of the case that, I think, contribute to the further development of compaction technology.

4.3 CORRELATIONS OF STANDARD AND MODIFIED PROCTOR DENSITIES

If I could, I would wave a magic wand and dissolve from everyone's memory the existence of standard Proctor densities, and decree that all subsequent specifications be based on an appropriate percentage of modified, as per the method I describe in Section 4.2.3. This method, by the way, has been endorsed by Ralph Peck, who reviewed the first edition of this book, as being similar to one he has used in his practice. Thus, I am proud to say that I am in pretty fancy company, however transiently (Andy Warhol was right!). I have no illusions that I will have a geotechnical volume done in my honor, but it would be nice if some young geotechnical engineer out there would dedicate one of his or her daily fill inspection reports to me . . . and maybe buy me a beer some day when we cross paths.

Having said that, I recognize that, realistically, SP and MP are here to stay, so here are some correlations, for what they're worth. Actually, I believe one genuine value of the comparisons is that they do emphasize again the very significant differences between standard and modified, and that's all to the good, and consistent with earlier commentary in this book.

Five soils were chosen as being reasonably representative of the soils of the sewer backfill. Four were gravelly sands, and one was a silty sand. For the four gravelly sands, 95% modified Proctor (MP) corresponded to about 103% standard Proctor (SP); this is an average value for the four soils. For the silty sand, 95% MP corresponded to about 100% SP, the somewhat smaller spread being sensible for the more uniformly graded soil.

An earlier correlation study yielded an average value, for 53 soils, of about 87% MP corresponding to 95% SP.

4.4 BEARING CAPACITIES OF PROCTOR DENSITIES

As developed previously, the void ratio of a soil can be expressed as

$$e = \frac{G_s \gamma_W V}{W_S} - 1$$

Rearranging gives

$$e = \frac{G_s \gamma_W}{W_S/V} - 1$$

But, W_S/V is the dry density γ_d. $G_s = 2.65$ (closely, for granular soils), and $\gamma_W =$ 62.4 pcf. Thus,

$$e = \frac{165.4}{\gamma_d} - 1$$

If one has the compaction curve for the soil in question, the computed value for the desired level of compaction can be substituted to obtain the void ratio of the soil. If, as before, one assumes that the loosest and densest void ratios of a granular soil are 0.8 and 0.5, respectively, a reasonable range for granular soils, the relative density of the soil can be calculated. Then, using Table 4.1, an approximate value of N can be obtained.

Finally, with this value of N, an approximate bearing capacity for the soil can be determined from Figure 4.2.

4.4.1 Standard Proctor

For the case study of Section 4.2.4, a SP moisture–density (compaction) curve for a soil of interest yielded a maximum value of 108.5 pcf. The void ratio for the soil at 95% SP is thus

$$e = \frac{165.4}{(.95)(108.5)} - 1$$

$$e = 0.604$$

And

$$D_R = \frac{0.8 - 0.604}{0.8 - 0.5}$$

$$D_R = 0.653$$

Interpolating from Table 4.1 yields a blow count:

$$N = 27$$

Using Figure 4.2 yields a bearing capacity of about 2.4 tsf, or 4800 psf.

4.4.2 Modified Proctor

Taking another soil from the case study, with a MP maximum of 119.7 pcf, and performing the identical steps yields a blow count over 50 (say 95% D_R), and a corresponding bearing capacity of about 5 tsf, or 10,000 psf.

4.4.3 Discussion

If the compaction requirements are specified at even higher levels than 95% MP (say 97%), as was done in this case study, then I would judge that the bearing capacity would increase to perhaps 10 tsf (20,000 psf), because, as inferred by Figure 4.4 and the related commentary, compaction at this level yields densities that approach 100% relative density.

As noted on the ordinate of Figure 4.2, the chart is based on the settlement criterion, inasmuch as deep-seated, classic bearing capacity (shear) failures are virtually unknown in granular soils (except possibly for very loose soils). What we would have for a soil compacted to 97% MP would be comparable to bedrock in bearing capacity, so 20,000 psf is a fair ballpark estimate.

For the residential construction of this case study, the bearing pressures would be a tiny fraction of this, perhaps a couple of hundred psf. To quote my first "boss" in geotechnical practice, Dave Greer, a retired preeminent engineer (to whom this book is dedicated), bearing capacities of residences are "usually unimportant."

Thus, the houses would have a factor of safety against bearing capacity failure of about 100. Since the usual safety factor is 3, I would say that the specifications for such compaction were perhaps a tad conservative. Indeed, I had the feeling that they were even a bit punitive, not to mention unnecessarily costly.

This case study is still another example of the need for a better understanding of compacted fill technology, both with respect to the importance of monitoring filling operations, and in writing sensible specifications. In this rèvealing case, the absence of the former led to protracted problems and huge and unnecessary costs.

4.5 CHANGING BORROW

One of the most vexing problems in fill control occurs on jobs where the texture of the borrow being brought to the site is changing with some frequency. It is not often that borrow areas are investigated by borings, except of course for very large jobs, and so the changes are typically unexpected. When such a change occurs, it seems that a compaction test should be performed on the "new" borrow. However, if the test is to be done in accordance with ASTM procedures, a period of approximately one week is required. (See Chapter 3 for a description of the sequence of test requirements.) If the borrow area is unusually heterogeneous, the problem can become acute and virtually unmanageable because of week-long delays for compaction testing each time the borrow changes unexpectedly.

4.5.1 The Compaction Data Book

To solve this problem, I recommend the development of a published standardized document that will enable fill control personnel to *predict* modified AASHTO density within minutes for any soil exhibiting texture that qualifies as load-bearing fill; TALB or select fill. In my suggested auxiliary classification system, a GS, an ECS, and (maybe) a gemicoss would qualify, the latter depending on job circumstances such as cost, available funds, and safety. I propose that the document evolve in two stages.

STAGE 1 I envision the stage 1 document as consisting of perhaps 100 grain size distribution curves encompassing the spectrum of acceptable textural requirements. These curves would be arranged in a standardized loose-leaf binder in logical sequence of incremental gradation changes. Included on each sheet would be the experimentally determined 100% modified AASHTO density, as determined in a carefully controlled applied research project, plus a listing of all index properties that are judged to affect significantly the compacted density of the soil. It is my judgment that these should include a uniformity coefficient = D_{60}/D_{10}; Hazen's effective size, D_{10}; percentage fines; the plasticity index of fines (PI); and angularity. Also, in some cases, specific gravity differences may be a factor. A proposed format is shown in Figure 4.5.

The procedure for using the method (stage 1) would entail the use of overlays, as follows:

1. On suitable transparent plastic (Mylar) overlay sheets, plot the grain size distribution curve for the site borrow. (Blank overlay sheets would be prepared in advance with scales identical to those of the standardized scale of the book. Time permitting, the grain size analysis could be completed in the laboratory. Alternatively, field personnel could generate the data by hand sieving.)

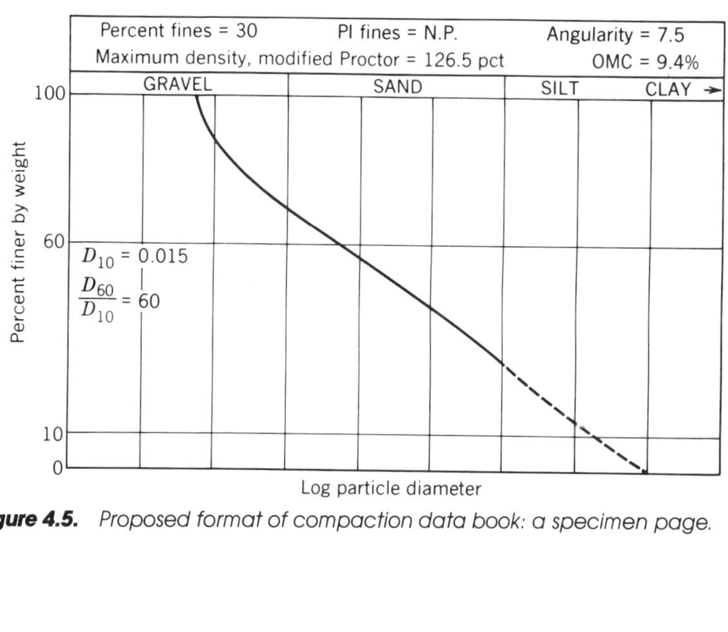

Figure 4.5. *Proposed format of compaction data book: a specimen page.*

2. Overlaying the plotted curve, search the book for the best fit. Note the 100% modified AASHTO density of the selected best fit.
3. To the extent judgment, experience, and confidence permits, make whatever corrections are indicated by any or all index property variations between the borrow and the selected best fit.

The last step in the stage 1 sequence is, as the saying goes, quite a mouthful. An inexperienced young person on first assignment could not be expected *at the present time* to make such judgments with any degree of confidence. I propose, therefore, that the stage 1 procedure would generally omit any attempts at corrections, except when arrived at by, or under the guidance of, an experienced specialist. The procedure should be used, initially, only on jobs where the borrow is changing with great (and frustrating) frequency. Because of the one-week typical delay that would be required for each compaction test for each "new" soil, there really is no better alternative than to use (generally) uncorrected best fits to determine target values, especially when one considers that, with the availability of the book, the process would yield an answer in a matter of minutes.

STAGE 2 I propose that stage 2 be the development of straightforward methods for determining the corrections noted in step 3 of stage 1. As an example of such a correction, it may be observed that well-graded soils have a greater potential for higher density than poorly graded soils, the former having successively smaller particles "available" for filling successively smaller void spaces. Thus, if the borrow soil has a numerically lower uniformity coefficient (poorer gradation) than the best fit, the density correction for this index variation would be negative. What the specific numerical value of the correction would be is part of the investigation of stage 2. Inasmuch as the determination of methods for assessing all corrections would require a major research effort or a series of investigations that could take many years to complete fully, I believe the sensible solution is to develop the method in the "stage approach" described. Actually, stage 2 would have a set of substages, in that when a method has been established for correcting any one of the index variations, it could be immediately incorporated as a refinement to the book.

SUMMARY OF RECOMMENDATIONS What I am suggesting and recommending is that the refinements could come one at a time, over a period of many years, or all at once. I believe that the latter could occur only if a major research project was funded and undertaken at some large university with a strong geotechnical program, involving a team of graduate students directed by a bright energetic professor with strong interest, experience, and expertise in geotechnical practice and applied research.

I offer the foregoing research ideas and also appeal to some agency or group of practitioners to consider funding the projects. Stage 1, the development of the compaction data book, without correction features, is, I believe, a relatively modest and worthwhile undertaking, particularly worthwhile because it "solves" the problem of changing borrow. Stage 2, a much more ambitious undertaking, if completed to a demonstrated high degree of reliability, would completely eliminate the need for compaction testing for all time and for all compacted fill jobs (admittedly perhaps too optimistic).

Chapter 5 contains additional commentary and recommendations regarding aspects and details of applied research and development with which I have been involved over a period of 34 years.

4.6 PROBLEMS EVOLVING FROM TRADITIONAL PRACTICE

Many of the problems in this category have already been presented, because they could not be logically separated from other more explicit problems, so this section serves partly as a summary. Included also, however, are some suggestions for solutions that I feel would be of benefit to all concerned parties in earthwork aspects of engineered construction.

4.6.1 The Roles of Nonspecialists

Case Studies 1 and 2 illustrated the role and related problems of two very important nonspecialists: the structural engineer and the earthwork contractor.

THE STRUCTURAL ENGINEER The basic cause of problems related to the structural engineer's functions is tied to the simple fact that the responsibility for structural drawings are logically those of the structural engineer, *including the foundation plan.* (See especially Case Study 2.) Included on the foundation plan are foundation notes. There is, therefore, a routinely created situation, frequently exacerbated by time and cost pressures, wherein the temptation exists for the structural engineer to write the foundation notes and any needed related (separate) specifications dealing with soil compaction.

My basic recommendation is this: Structural engineers should not write foundation notes or other soil compaction specifications independent of advice from soil specialists. (I don't design high-rise buildings!) Alternatively, the structural engineer should undertake a reasonably intensive study of soil compaction and fill control so that, in effect, he or she becomes a specialist in this one very important area of geotechnical practice. One step toward achieving the latter objective would be the inclusion of substantial coverage of these topics in undergraduate courses in soil mechanics, and that is my secondary recommendation. This procedure would provide a partial solution for graduating engineers of the future, and pertinent parts of this book will, hopefully, serve the needs of those structural engineers already practicing.

I have always maintained that good engineering education should include the topic "when to consult a specialist," or "knowing what you know."

CONSTRUCTION ENGINEERS AND CONTRACTORS The foregoing commentary and recommendations will be helpful also to construction engineers and contractors, since success in bidding, planning, and supervision are inextricably linked to the ability to assess and solve earthwork problems, including interpretation of geotechnical data, boring logs, and excavation and compaction specifications.

ARCHITECTS AND OWNERS Although considerably removed from the details of earthwork operations, the necessary interactions between the architect and the structural engineer, and between the architect and the owner, would suggest that general knowledgeability would be an important asset. When structures crack, everybody gets to see it, including the owner. Chapter 2, Avoiding Costly Blunders, is recommended for minimum knowledgeability.

4.6.2 Fill Inspection Personnel

The problem here is that, traditionally, earthwork inspection has almost invariably been entrusted to rookies; such activity is usually one of the first assignments a young engineering graduate receives. Since undergraduate curricula provide inadequate information (in my view), the young fill inspector learns on the job, usually helped by periodic advice from the experienced project engineer, but more often left to muddle through in a trial-and-error fashion. Ironically, the better young engineer is promoted to more important activities in the office about the time he or she becomes proficient at fill inspection. Thus, there exists a vicious cycle, the result of which is cracked and failed structures.

Fortunately, there is presently evolving an excellent solution to this problem, independent of any efforts of mine: the emergence of degree programs in engineering technology (Bachelor of Science in Engineering Technology). It may take some time to graduate enough technology specialists to effect a permanent solution to this problem, but the process is under way. In the meantime, it is recommended that young engineers, engineering technologists, and those presently involved directly with fill inspection become thoroughly familiar with material contained in Chapters 6–8. Chapters 7 and 8, Compaction Specifications and Fill Control Procedures, are, of course, particularly applicable.

4.6.3 Relationships Among Concerned Parties: Contractor, Inspector, Engineer

The fill inspector, above all others, is in a most delicate position. On the one hand, first and foremost is the responsibility to the owner: to ensure that the work is executed in compliance with specifications. On the other hand is the desire to get along amicably with the contractor who is executing the work. The contractor, or the contractor's earthwork foreman, is the person with whom the inspector must interact on a daily basis. In a scenario that is certain to occur, the foreman has about 30 years of experience, knows nor cares little about field density testing, and regards all 22-year-old inspectors as novices. Add to this the fact that said novice knows little or nothing about filling operations, and the stage is set for some interesting scenes!

Such situations can and do occur, and can be aggravated immensely when an unscrupulous contractor is involved. In Chapter 8, I describe just such a case history—a real horror story.

The engineer (i.e., the project engineer) has accumulated the experience and professional maturity to handle effectively the situations that arise in the field, but the problem is that he or she is not in the field on a regular, daily basis. The typical project engineer may be involved in a supervisory capacity with a number of small projects, and thus is in contact with a fill inspector on one of these projects only on a periodic or as-needed basis. This may include contacts by phone, by periodic written reports (perhaps also sent to the client), and by the engineer's occasional visits to the site.

In Chapter 8, Fill Control Procedures, I offer a series of suggestions on how a fill inspector might deal with various problems that may (and do) arise in relationships with all concerned parties.

4.7 COST AND TIME PRESSURES: A SUMMARY

"Time is money." "I need it tomorrow." I wish I had a dollar for each time these phrases have been uttered in the frenetic world of engineered construction. The complicating factors in earthwork construction are (1) the inherent uncertainty regarding subsurface conditions (with or without borings) and (2) the weather and its particular pronounced effects on soils and site conditions.

I have alluded to specific pressures associated with compacted-fill construction. These are caused by

1. The reluctance or procrastination in calling in a soils specialist, especially for large jobs. (This is often a cost *and* a time pressure.) It is suggested, however, that the failure to do so will often lead to greater costs in the long run, not to mention the costs measured in grief, frustration, and embarrassment. Potential legal consequences, in increasing instances, are an added factor.
2. The need to decide on the degree of compaction of a fill for a wide variety of loading conditions.
3. The particularly vexing problem of changing borrow.

The foregoing are the easy problems to solve, since they are largely technical in nature and thus susceptible to technical solutions. The much more difficult problems are those that are rooted in traditional practice and human interactions. Chapter 8 deals more extensively with these problems. Unfortunately I have no pat answers, as I believe there are none.

4.8 GLOSSARY

Flat Dilatometer Test (DMT) The Marchetti Dilatometer Test, which consists of a steel soil penetrometer device pushed vertically from the surface and stopped at preselected depths to perform a test at each depth. The device is 97 mm wide, 18 mm thick, with a 60-mm expandable diaphragm on one face. Each test consists of measuring expansion pressures, and provides information for geotechnical engineering exploration and design. (Schmertmann, 1993, personal correspondence).

In situ A Latin phrase meaning "in its natural or original position." Thus the *in situ* moisture content, unconfined compressive strength, etc., would refer to the natural moisture content, the undisturbed strength, etc.

Standard penetration test (SPT) An important field sampling test, wherein a sampler (called a split spoon sampler) is driven into the ground in a standard fashion (140 lb hammer dropping 30 in.). During the driving, blows of the hammer are counted for two successive 6-in. intervals, yielding the blow count, an indicator of soil density (*in situ*).

Tremie An arrangement of linked tubes, with a wide funnel at the top, used for pouring concrete into relatively inaccessible excavations; used to avoid free-fall (and thus segregation) of the concrete.

Uniformity coefficient The ratio of the D_{60} size to the D_{10} size, thus an index of gradation of a soil (most meaningful for a granular soil). The larger the number, the wider the gradation. Obtained from a grain size distribution curve. See Section 5.1.3.

Void ratio The ratio of the volume of voids to the volume of solids, expressed as a decimal. Thus, a soil with a high void ratio has a low density, and vice versa.

CHAPTER 5

Applied Research and Development

Over the years of my interest in compacted fill technology, stimulated by early practical exposure as a rookie geotechnical engineer, I have initiated investigations into a wide variety of factors that I felt needed more study. Some of the work has been completed to the point of publication and patents, while other studies are in various stages of early development. None of the work was sponsored, funded research. The information that has been collected, then, was largely the work of former students under my supervision at the Newark College of Engineering of New Jersey Institute of Technology.

I will not present details of this work but rather only the highlights, the purpose of which is to stimulate interest in further development to the stage of practical applications, notably with respect to the proposed compaction data book. Factors that are suggested for further study are grouped as follows:

1. Effects of specific index property variations
2. Artificial fills
3. Low-level geothermal heating and cooling, energy conservation
4. Waste materials as fills
5. Effects of mechanical laboratory compactors
6. Density gradients
7. Geostick correlations
8. Percent compaction (specifications) for clay fills

5.1 EFFECTS OF SPECIFIC INDEX PROPERTY VARIATIONS

To develop the compaction data book to its fullest (stage 2, Chapter 4), it will be necessary to determine the effects of specific index property variations to a degree that will allow for numerical corrections for each variation. (Recall that "variation" refers to each variation from the "best fit" of the compaction data book. See stage 1, Chapter 4.)

5.1.1 Percent and PI of Fines

Authorities recognize that the amount of fines in an otherwise predominantly granular soil (EGS) will affect the engineering characteristics of the soil. Not as generally recognized, at least explicitly, is that the *plasticity* of the fines (PI) will also affect the engineering behavior of the soil. For example, some of my earliest work involved experimenting with varying proportions of granulars (+200) and fines (−200). Using commercially available powdered kaolinite, a white clay, as fines, I found at least 15% (by weight) was required to impart cohesiveness or binding qualities to a standard granular (percentages retained on the 40, 100, and 200 sieves were 67, 31, and 2, respectively.) The liquid and plastic limits of the kaolinite were 60 and 30, respectively, yielding a PI of 30. The soil mixtures thus prepared were compacted at about 12% moisture in a Harvard miniature compaction mold, extruded, and tested for unconfined compression strength. (For descriptions of equipment and testing procedures, see Lambe, 1951, p. 44.) Ten percent kaolinite merely served to coat the granular particles. As a result, the 90–10 mixture retained free-draining qualities, and compaction was fruitless in that the added moisture oozed from the bottom of the mold upon application of energy. At 80–20 the results were, as expected, better: less oozing and an extruded sample of sufficient cohesion to run an unconfined compression test.

Using these preliminary findings as a base, a research program was initiated wherein student groups were assigned 70–30 mixtures with the 30% fines having varying degrees of plasticity: marble dust (nonplastic), kaolinite, a modeling clay, and bentonite (drilling mud). Cartons full of data on this and other (later) projects have been accumulated but have not been carefully evaluated, nor would its inclusion here be compatible with the purpose of this book. However, a few observations and suggestions are offered regarding effects of fines in general.

NONPLASTIC FINES An ideal granular soil (one that is permanently stable) is dense, well-graded, free-draining, angular, and laterally contained. (The last requirement may be illustrated by noting that excavation adjacent to and below a footing would cause the undermining of that footing by lateral flow of the cohesionless, granular soil from beneath the footing.)

The existence of nonplastic fines (clean silts) in a granular soil will create a potential for higher density, since the void spaces the silts would occupy would otherwise be filled with air or water. However, if the percentage is high enough, these fines will clog the voids and thus render the soil marginally draining rather than free draining. The detrimental effects of changed drainage characteristics is, I judge, a bad trade-off. These effects include susceptibility to frost action, potentially damaging seepage pressures, and liquefaction (a sudden and complete loss of strength, resulting in collapse of supported structures*). From a compaction standpoint, such a soil would also be difficult to compact because of its moisture sensitivity. Finally, the fines would impart no strength to the soil by what is often called "desirable cohesive binder," because of its nonplastic, cohesionless nature.

From this, one can say that, of the four recognized soil constituents (gravel, sand, silt, clay), silt is the only one that has nothing going for it. (While this is true for *load-bearing* fills, it is *not* true for an important class of *un*compacted fill, select fill for septic system mounds. This important fill is considered in Chapter 9.)

PLASTIC FINES As noted, these fines *do* impart binding qualities to the soil mass. If the fines are present in sufficient quantities and degrees of plasticity, however, the potential problems can far outweigh, or negate, any desirable binder benefits. Principal problems are potential compressibility or expansiveness, and imperviousness. The latter characteristic entails additional problems of seepage pressures (including uplift), and excess porewater pressures (reducing shear strength). Clays are also compaction-energy sensitive, and compacted embankments of high clay content are highly anisotropic, exhibiting markedly different strength and seepage characteristics in different directions.

Since compaction is inherently a lubrication process (Figure 3.1), the interaction of water with a fill containing plastic fines (or one that is 100% clay) may be thought of as "shared water." As described previously, clay particles of different mineralogies may have markedly different affinities for water. The water that attaches to the clay particle surfaces (actually called "bound water" by many authorities) has many of its properties changed by complex electrochemical activity, a fundamental cause of plasticity. One of these changes renders the bound water more viscous than normal water, and thus less capable of providing the lubrication necessary for compaction. Thus, water added to a dry clay soil can be thought of as being divided into bound water and lubricating water, the proportions depending on how much the particular clay mineral requires for bound water satiation. For highly plastic clays such as bentonite (liquid limit about 400%), one would have to satisfy this bound water demand before any lubrication can take place for effective compaction. This reasoning makes it easy to see why optimum moisture contents for fat clays are so high compared to lean clays such as kaolinite (Figure 3.4).

*Contractors and equipment operators should be aware of the very real danger of equipment (and operators!) being swallowed up with very little warning if an overly vigorous attempt is made to compact thick lifts of loose, saturated sands with heavy vibratory compactors, particularly when dealing with hydraulic fills (Sowers, 1970, p. 233).

COMPACTION DATA BOOK CORRECTIONS For purposes of assessing the effects of percent and PI variations, it would seem reasonable to conclude the following:

1. The density corrections for a PI variation would be inversely related. That is, if the PI of the site borrow fines is higher than the best fit, its effect would be to lower the compacted density of the borrow, all other factors being equal.
2. The density correction for a percentage fines variation would also be inversely related; the more fines, the lower potential density.
3. Optimum moisture content corrections would be directly related to PI variations. Thus, if the PI of the site borrow fines is higher than the best fit, the optimum moisture content of the site borrow would be higher.

5.1.2 Hazen's Effective Size: D_{10}

The D_{10} size was postulated by Allen Hazen to be the size that most significantly affects the seepage qualities of a soil. Hazen's work dealt principally with sand filters in water treatment plants (i.e., sanitary engineering). He probably reasoned, quite logically, that the sizes of the smaller void spaces would govern drainage qualities, and that these void spaces would be related to comparable particle sizes. The 10% size (meaning that 10% of the soil is finer than that size—see Figure 4.5) was chosen for study, and an empirical equation was developed relating permeability to the D_{10} size:

$$k = cD_{10}^{2}$$

where k is the permeability coefficient (in millimeters per second), c is a constant between 10 and 15, and D_{10} is Hazen's effective size in millimeters (Sowers, 1979, p. 96). The equation is principally applicable to clean sands and is usually considered to give only an indication of the order of magnitude of the permeability.

A preliminary investigation has been made to determine the relationship between D_{10} and the compaction characteristics of soils bordering between freely draining soils and marginally draining soils. As illustrated in Figure 3.4, select fills (freely draining) do not exhibit a discrete optimum moisture content, and, therefore, it makes no sense to attempt a standard compaction test on such a soil. The problem is, of course, to have a way of identifying such a soil quickly by some easily determined index property. It was the purpose of my investigation to determine the approximate D_{10} size that identifies a select fill as such, and that is one of my suggestions for future applied research and development. (As noted in Chapter 1, Section 1.4, select fills are sometimes defined as soils containing not more than 12% passing the No. 200 sieve. This would serve as a good starting point for the investigation I propose.)

5.1.3 Uniformity Coefficient: D_{60}/D_{10}

A well-graded soil is one that contains successively smaller particles filling successively smaller void spaces, thus creating a potential for greater density. The uniformity coefficient D_{60}/D_{10} is a number or index that reflects the gradation of the soil (see Figure 4.5). Clearly the larger the uniformity coefficient, the wider the gradation. Thus, if the site borrow has a greater uniformity coefficient than the best fit, the compacted density of the site borrow should be greater.

No research has been done (to my knowledge) to determine a specific relationship between uniformity coefficient and compacted density. The uniformity coefficient is a standard index property in the Unified Soil Classification System. Thus, the research I propose would have immediate and easy applicability in practice.

5.1.4 Angularity

Geologic processes influence greatly the angularity of predominantly granular soils, particularly the abrasion effects during transport by water or wind (alluvial and aeolian soils, respectively). In general, highly angular soils, if dense, are extremely strong and stable because of the interlocking of particles. Angular particles also exhibit rough surface texture, like sandpaper, so that particles interlock also in a surface-to-surface fashion. Thus, interlocking may be thought of as occurring on a macroscopic and microscopic level. Means and Parcher (1963) contend that the angularity of granular soils has a much greater effect on the soil's behavior than the paucity of investigations suggests, and I concur.

One of the first steps in such an investigation would be the development of a practical method of assessing a soil's angularity quantitatively analogous to the hardness number of minerals. Thus, a soil might have an angularity rating between, say, 1 and 10 for spherical and pyramidal shapes, respectively. Once such a rating method is established, it would be possible to study angularity effects on compaction and other engineering properties, such as shear strength and compressibility.

I have supervised some work along these lines. As a first step, the literature was searched to determine if any attempts had been made by other investigators to quantify and study angularity. As I suspected, little if any such work had been done. Indeed, all textbooks seem to classify granular soils as angular, subangular, subrounded, and rounded, and let it go at that. Angularity effects appears to be a neglected area of study.

5.1.5 Specific Gravity

Means and Parcher (1963) suggest that the specific gravity of granular soils, which would be composed primarily of quartzitic minerals, can be taken as

2.65 for most practical applications—that of clay minerals somewhat higher, about 2.70. It is evident, however, that some instances will arise where "peculiar" mineralogy will dictate that the assumption of a specific gravity of 2.65, 2.70, or an intermediate value for a mixed soil, will introduce unacceptable error. Fortunately, with experience, one can easily identify situations where specific gravity variations may be significant. There are two ways in which this can be done. First, the soil feels heavy (or light). It is surprising how this simple tactile technique allows one to detect significant specific gravity variations. For example, the specific gravity of cement is 3.1, and even an inexperienced person can easily detect the "heaviness" of a handful of dry cement compared to, say, a handful of powdered bentonite (drilling mud). Try it! Second, one can develop an intuition for the "correct" maximum compacted density, at a given level, as suggested by Figure 3.4. When the answer is suspect, it may be the result of significant deviation of the specific gravity from the usual 2.65–2.70 range.

When such deviation is suspected the specific gravity should be determined. I would, however, offer a note of caution: Be sure to have the work done by a company whose laboratory work you have learned to accept as accurate. It is my opinion that the classical, direct method of determining the specific gravity of soil minerals is unusually susceptible to experimental error, so much so that assuming an answer is often more accurate than trusting an experimental result, particularly when experienced persons apply their judgment to the specific case.

I have seen data that suggest that only the very best laboratories can perform this "simple" test satisfactorily. Sometime around 1960, the American Council of Independent Laboratories conducted a Standard Soil Samples Program, which involved sending soil samples to 99 participating laboratories for certain routine soils tests, including specific gravity determinations and modified AASHTO compaction.

Three "umpire" laboratories, presumably chosen for their reputations and prestige, were chosen to do the same tests. Their results established the control (i.e., correct) values.

Following are the maximum, control, and minimum values reported for a CH soil (clay of high plasticity, Unified System) (Hirschfeld, 1965):

	Maximum	Control	Minimum
Specific gravity	2.79	2.70	2.21
Modified AASHTO	123.9	114.0	105.4

And, mind you, since these were participating laboratories, they undoubtedly put forth their very best efforts.

To some degree, this cautionary note pertains to any laboratory testing you may need. Where very large jobs are involved, "second opinions" may be warranted.

5.2 ARTIFICIAL FILLS

In recent years, the use of artificial fills has become more common for a variety of reasons. In and around urban areas, the availability of large quantities of fills of suitable texture has become a problem, and many of acceptable texture are often found to be contaminated. Environmental regulations pertaining to filling operations have also made the use of "normal" fills more costly and problematic.

Artificial fills are here defined as those that are manufactured or processed in some way, and include foam plastics such as Styrofoam and Styropor, Elastizell, Geocell, and Solite. Most of these products are proprietary. Names and addresses of the manufactures are included for interested readers, but inclusion of descriptive materials are not intended to endorse any particular product.

An additional feature of all of the materials described is that they are all lighter in unit weight than inorganic, natural fills, and so a weight credit is associated with their use. When dealing with construction over weak soils, this advantage can be of paramount importance.

5.2.1 Weight-Credit Foundation Construction Using Foam Plastic as Fill

If a hole is dug and the material that is removed is immediately used to fill the hole, the state of stress below is unchanged. It follows that if a lightweight backfill is used, there will accrue a weight credit for a proposed structure. In an extreme case, if the weight of the backfill plus the structure is equal to the weight of the soil removed, then no settlement of the structure can occur. This principle is used extensively in Mexico City to produce so-called floating foundations on the highly compressible volcanic clays underlying the region. In less dramatic cases, lightweight backfills (e.g., cinder fills) may be used to effect a lesser weight credit, thus reducing the net stress increase on compressible materials below. Apparently because of the scarcity of natural lightweight aggregates and the consequent infrequency of fortuitous circumstances that might dictate it use, the application of the technique has not been extensive.

I propose the use of foam plastic as backfill in certain special cases. Technology exists or can easily be developed for on-site production of the foam plastic, or the material can be precast for installation at the site.

ILLUSTRATION OF WEIGHT CREDIT Consider a site that is overlaid by compressible material of soft consistency, for example, a swamp muck.

Unit weight of muck	100 pcf
Depth of excavation	15 ft
Stress release	1500 psf

Foam density	3 pcf (typical)
Stress addition	45 psf
Stress credit	1455 psf

It is common that in areas where soils of such marginal supporting capacity exist, the typical methods of soil stabilization (e.g., preconsolidation with or without sand drains or the use of compacted sand backfill) are suitable for developing an improved condition to allow building pressures of about 500 psf, corresponding to one-story light industrial buildings or office buildings. It is seen that the stress credit in this example is about three times this value. And this trebled allowable building pressure could theoretically be applied without causing any settlement.

The compressive strength of a foam that has been used for such an application (Styrofoam HI) is 35 psi. Thus, the building pressure of 1425 psf (±10 psi) would stress the plastic to a level of less than one-third of its compressive strength.

A perhaps oversimplified conclusion is that a three- or four-story building would now be possible where only one-story construction would be technically feasible using standard techniques.

FOAM MATERIALS The only foam plastic that is known to have been used in weight-credit engineered construction is a (solid) precast extruded polystyrene foam designated Styrofoam HI, one of approximately 12 Styrofoam materials made by the Dow Chemical Company. Styrofoam HI was used on the construction of the Pickford Bridge in Michigan (Coleman, 1974).

To illustrate the broad potential for the general use of foam plastics (precast and blown-in-place) in weight-credit construction, consider the following quotation from *Modern Plastics Encyclopedia* (1968), "Urethane Foams," p. 346:

> [F]oams may be produced which have densities ranging from less than one pcf to about 70 pcf, with an almost limitless range of chemical and mechanical properties.

In light of the above, it would seem feasible that for very large jobs, or very special circumstances, one could justify the expense of producing a foam plastic of special formulation to suit the particular use. In addition, one could use combinations of existing products (i.e., Styrofoam products of different properties) to effect a design. As an example, one could use a high-quality Styrofoam immediately beneath a footing where stress levels are relatively high and a lower-quality (and presumably cheaper) foam plastic below. This would be analogous to the manner in which pavement systems are designed.

Table 5.1 lists technical data for the specific Styrofoam product that was used on the Pickford Bridge, where a wedge-shaped block of "bundles" of Styrofoam HI-35 boards were installed. Figure 5.1 illustrates the construction

TABLE 5.1 Properties of Styrofoam HI-35 (from Dow Chemical Brochure)

Property	Test Data	Test Method
Compressive strength at 5% deflection	35 psi (2.5 tsf)	ASTM D1621-59 T
Water absorption	0.25% (by volume)	ASTM C-272-53
Density	2.5 lb/ft^3	

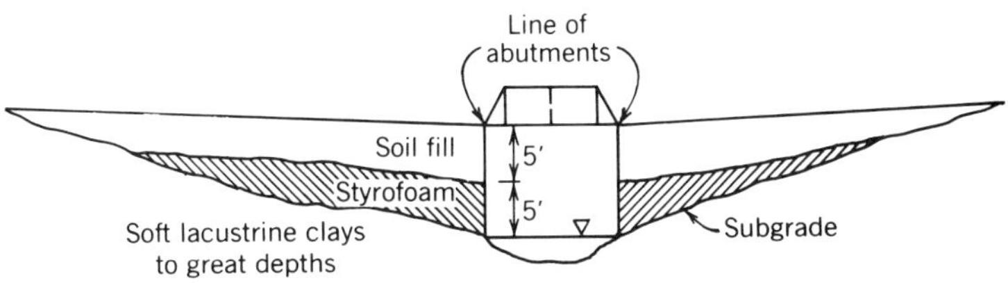

Figure 5.1. *The Pickford Bridge crossing, Pickford, MI.*

schematically. The soft clays were so weak that the abutments (prior to weight-credit construction) were tied to trees for added stability! With the use of Styrofoam fill, the maximum pressure (at the rear of the abutment walls) was reduced from about 1200 psf to about 612 psf (5 ft of soil fill at 120 + 5 ft of Styrofoam at 2.5). The soft clays were able to sustain this reduced pressure without detrimental settlements.

The 5 ft of soil fill was for two purposes: to pin down the foam plastic to avoid possible flotation and to minimize the amount of relatively expensive foam plastic ($40 per yard).

I was informed by the Dow engineer involved with the project that the subgrade was so soft that one would sink to one's knees when attempting to traverse the area with normal footwear. The bundles of Styrofoam were placed directly on the untreated subgrade, and two men placed one side in one morning.

A sheet of polystyrene was placed on top of the Styrofoam to protect against the possibility of deterioration of the plastic by spillage of gasoline in the unlikely event of a tank-truck crash at the approaches to the bridge.

The water absorption characteristic of the material used for a weight-credit application is important, inasmuch as water pickup will result in weight increase. To quote a prominent engineer whom I consulted: "I hate to bury voids" (Johnston, conversation, 1976). As noted, the water absorption of Styrofoam HI is virtually negligible (0.25% by volume).

Table 5.2 contains strength–weight ratios that illustrate by contrast this unusual property of rigid foam plastics.

PERMANENCY, DURABILITY The most convincing evidence of permanency and durability are those instances of field usage.

TABLE 5.2 Strength-Weight Ratios

Material	Compressive Strength tsf / Unit Weight (pcf)
Gravel (exceptionally compact)[a]	$\frac{10}{130} = 0.08$
Bedrock (massive, sound)[a]	$\frac{100}{170} = 0.59$
Styrofoam HI-35[b]	$\frac{2.5}{2.5} = 1.00$
Reinforced concrete	$\frac{217}{150} = 1.45$
Styrofoam HD-300[b]	$\frac{8.64}{3.3} = 2.70$

[a] Various building codes; typical.
[b] Dow Chemical Company.

The Pickford Bridge This is the only known instance of large-volume use of foam plastic in a weight-credit construction application.* As reported to me orally, there was an immediate settlement of about 2 in.—apparently an adjustment of the stacked bundles of Styrofoam—but no subsequent movement of significance has been observed since.

Highway Insulation Since 1962, Styrofoam HI has been used in at least 39 installations as insulation for highway pavement systems. In most cases, the amount of plastic used has been about 1 to 3 in. (typically placed directly on the subgrade). Samples of the foam taken from various highways after several years of service have shown very little moisture pickup (Williams, 1968).

Structural Performance The structural performance of a highway insulated with Styrofoam HI was investigated by personnel of the Maine State Highway Commission (1966). Deflections were of the order of 0.02 in. in the insulated sections and were well below those of the control section during the spring thaw, the latter reaching a maximum of 0.03 in.

Utility Installations Thirteen utility installations were insulated with Styrofoam between 1964 and 1966. These included sewer lines and water lines. No case of a pipeline application involving the purposeful use of foam plastic *for weight credit* is known.

*Since publication of the 1st edition, a large number of projects using foam plastic as lightweight fills has been discovered, and are described in Section 5.2.1, following.

Pole Installations A foam plastic compound called Poleset is being used to install poles in the ground. The injected compound is used in lieu of the usual technique of compacted soil backfill. Users, notably power companies, report in their literature that the material is stronger than the *in situ* soil, based upon pulling tests in the field. In such a usage, weight credit is not a factor. Apparently, the scarcity of select borrow, the neatness of the operation, and reduced labor costs combine to make the technique workable and economically feasible.

Laboratory Tests Accelerated laboratory tests such as freeze–thaw cycling and soak tests have shown very little increase in water pickup (Williams, 1968).

COST AND COST ANALYSES The in-place cost of Styrofoam HI at the Pickford Bridge was $40/yd. This was high compared to the cost of compacted fill in-place (perhaps $8/yd). However, a comparison of unit costs is not usually valid in appraising the feasibility of the weight-credit method. Instead, I suggest the following approaches.

Real Estate Context Perhaps the best context in which to gauge the worth and economic feasibility of a particular application is to compare the cost of stabilization (or *improved* stabilization) with that of the real estate value. For example, experts have estimated the value of lower Manhattan real estate at $400–500 per square foot (about $16 million per acre). The cost of placing Battery Park fill is estimated at about $14–15/ft^2, with a $40 upper limit when utilities and other costs are considered. Thus, it is clear that "making" this new ground (93 acres) was a tremendously sound investment.

In areas such as the New Jersey meadows, where the land is also inherently valuable (in a real estate context) because of its geography, but is undeveloped because of geologic (foundation) shortcomings, a similar type of cost comparison would be valid. Thus, if a foundation construction technique can yield a three- or four-story capability as opposed to the present one-story capability, the client gets three or four times as much floor space.

Total Cost Context In many cases, the best approach is to look at the bottom line or total cost of a job and compare that cost to other "traditional" construction methods. A Staten Island highway project demonstrates this approach. (See section following.)

Possibility Context In some instances, it may develop that no construction is possible without exceptional weight credits afforded by the use of foam plastic fill.

HYPOTHETICAL CASE HISTORIES To illustrate further the application of the weight-credit technique, a number of hypothetical case histories follow. Some are general and merely conceptual; but some suggested applications are for the solution of specific, real problems in foundation construction.

Route Construction in General If a route of any type (highway, railroad, pipeline) is to be constructed between two points, A and B, and a large area of marginal soil exists between these points, the usual solution would be to go around the area (see Figure 5.2). With the dramatic weight credit that can be attained by the use of foam plastic (as an artificial subbase for a highway, for example) it is conceivable that the marginal soil might be crossed. The length of the route would thus be shortened and the resultant savings might make the straight route economically superior. Possibly, the loads imposed by rail equipment would be so high as to preclude its application to railroads, but highway and pipeline loadings seem within the scope of application. When it is considered that interstate systems cost in excess of $1 million/mile (depending, of course, on cuts, fills, locale, etc.), the cost of the plastic foam and its installation could be justified.

Highways Healy (1975) cites a highway project that illustrates both the method and the manner of appraising in a total cost context:

> A recent example of a case where foam plastic backfill would have provided a better and less expensive solution to a soil problem was the construction of a major arterial highway in Staten Island, New York. The author was directly involved with this project which bisected a tidal marsh. The method specified in the contract for subgrade stabilization was to excavate unsuitable material and backfill with compacted 1½ inch broken stone. The purpose of using 1½ inch broken stone instead of sand fill was to have larger voids, thereby generating a larger weight credit. The contract unit price for stabilization was $33.00 per cubic yard—$13.00 for excavation and $20.00 for in-place 1½ inch broken stone fill.
>
> The subgrade was so poor that it was necessary to excavate as much as 10 feet in certain areas. The final excavation quantity for stabilization was 83,350 cubic yards which generated a total stabilization cost of $2,750,550.00. However, if foam plastic was utilized as backfill for the purpose of gaining weight credit, as

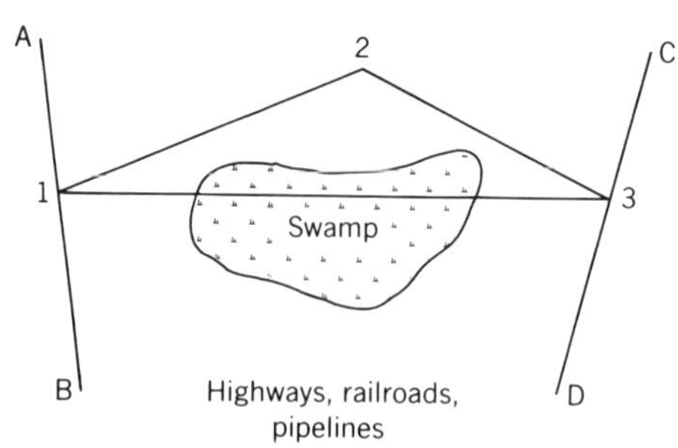

Figure 5.2. *Route locations.*

was suggested to the New York City Highway Department during the early stages of the project, a substantial savings would have been realized. Only 10% of the final excavation quantity (8,335 cubic yards) would have been required to achieve an equivalent weight credit. The total cost for stabilization if polyurethane was utilized would have been approximately $441,755.00 (8,335 cubic yards at $13.00 per cubic yard for excavation plus $40.00 per cubic yard for foam plastic backfill).

Therefore, it would seem that a savings of about 2.31 million dollars could have been achieved if foam plastic backfill was used. In addition, foam plastic probably would have produced a better overall project.

Pipelines The use of foam plastic as an injected backfill under and around pipelines is thought to be practical (see Figure 5.3). Envisioned is a type of wheel-mounted machine that would move in a straddling fashion along the trench, producing and injecting plastic foam into the trench. This application might be particularly appropriate in large, congested cities for two reasons. First, the existing pavement would provide the needed support for the machinery, and second, the trucking of soil fill to the site (through congested city streets) could be eliminated.

In the case of fluid in a pipeline that is designed for gravity flow, it is apparent that the size of the pipe wall can be reduced to that of essentially a form since the only reason for a pipe of any substantial strength in such a case is to withstand the pressure of the backfill and any live loads. Since the pressure exerted by a foam plastic fill would be negligibly small, the required strength of the pipe would be governed principally by live loads. In fact, it is speculated that a slip-form of some type could eliminate the need for a pipe; the slip-form could be moved forward after the plastic foam has created a conduit through which the fluid would flow.

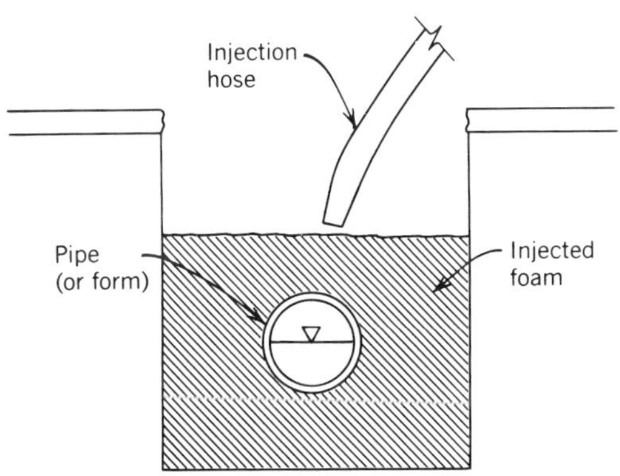

Figure 5.3. *Trenches.*

Thus, it is conceivable that, for the case of urban pipeline construction, the costs of soil fill, its transportation, and the costs of the pipe (gravity flow) could be reduced or even eliminated (i.e., weighed against the costs of the foam plastic weight-credit approach).

Because of the heat insulating properties of foam plastics, its use to encase the oil pipelines through permafrost is suggested. Where needed, weight credit and insulation could be achieved at the same time.

A Pile Load-Test Program I worked on a pile load-test program in which steel H-piles were driven to bedrock to depths of about 70 ft, with specifications for equipment and driving to produce a design pile of 175 tons (see Figure 5.4).

Above the bedrock were thick deposits of soft clay. Finished first-floor grade was at a level several feet above existing grade; and the design called for the space between existing grade and the slab to be a fill material. The high floor loads dictated a structural floor slab, so the only reason for the soil was to fill the space under the building. The weight of this fill would, however, cause the consolidation of the thick, compressible clay below, and it was estimated that the drag on each pile caused by the settling clay would be about 50 tons (negative skin friction).

The use of a foam plastic fill could have eliminated the substantial effect of drag on the piles, and smaller piles could have been used to support the building-imposed pile loading of 125 tons.

The Healy Pile (Figure 5.5) Another excerpt from Healy (1975) follows:

> The weight credit obtained if polyurethane foam plastic was substituted for concrete in piles would be substantial. The pile bearing capacity could then be increased by the resulting weight credit.

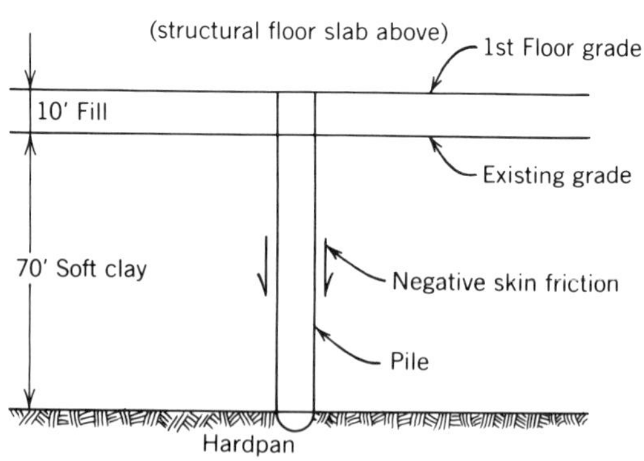

Figure 5.4. *Pile foundations.*

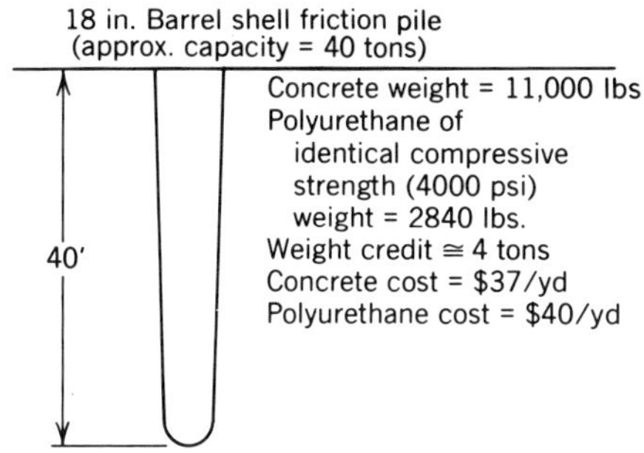

Figure 5.5. *The Healy pile.*

For example, an average 40 ton capacity, 18 inch diameter, barrel shell friction pile which would probably be driven to a depth of about 40 feet, would have a volume of approximately 71 cubic feet. If concrete was utilized to fill the pile, it would weigh approximately 11,000 pounds. On the other hand, polyurethane plastic, which weighs 40 pounds per cubic foot for an equivalent compressive strength (4000 psi), would yield a pile weight of 2,840 pounds; weight credit, about 4 tons. As a result the capacity of the pile could now be considered as 44 tons, an increase of 10%.

In general, the weight credit generated by substituting foam plastic in any deep foundation system which utilizes concrete, would produce significant increases in allowable bearing capacity. In addition, it should be noted that the cost of materials are approximately equal—4,000 pounds per square inch concrete costs about $37.00 per cubic yard and 4,000 psi polyurethane foam costs about $40.00 per cubic yard.

A Grade-Separation Case Study (Figure 5.6) I became aware of plans to effect a grade separation at an important intersection in a major eastern city. In discussions with the chief soils engineer it was learned that the principal complication of the design was that a subway ran directly beneath the inter-

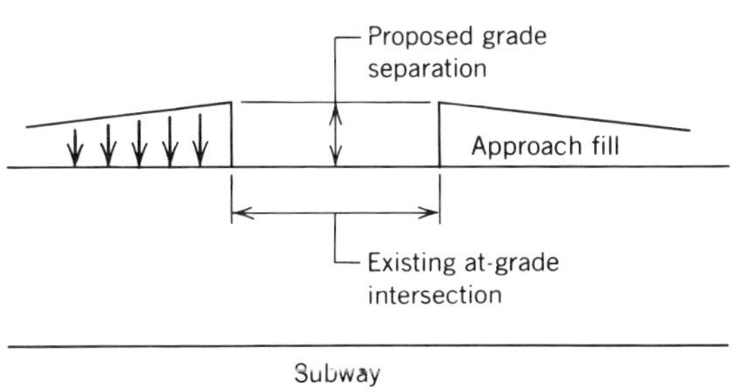

Figure 5.6. *A grade separation.*

section at a relatively shallow depth. And the subway at that point is presently loaded "about to its limit," so the approach fills necessary to make the grade separation create a problem of overstressing the structural elements of the subway system.

The solution was to excavate some of the material above the subway and backfill to the planned elevated grade with lightweight fill. I inspected the fill and it appeared to be some form of ash material containing considerable black silty material, a rather poor fill by normal standards (poor drainage, difficult compaction, probably frost susceptible), but its unit weight was about 65–70 pcf. Thus, a weight credit would be attained and apparently the material was acceptable for the solution of this special problem.

The use of a foam plastic approach fill (at least in part) could have provided a much greater weight credit and would have the advantages of ease of construction and elimination of drainage and frost problems.

The Catskill Aqueduct Case Study (Figure 5.7) Within a 48-mile section of the Catskill Aqueduct System there are approximately 15 miles of pressure tunnels. This construction was used to cross filled valleys where no suitable material was available to support other kinds of construction. In general, the depth of these pressure tunnels was governed by the depth of preglacial gorges that contained glacial drift and other recent unconsolidated alluvium. The deepest pressure tunnel was for the Hudson River crossing; it was founded at a depth of 1111 ft below the level of the river. The length of the tunnel at this depth is over 3000 ft.

No cost figures are available for the construction of these pressure tunnels, but the figures must be impressive for such an effort. Had the technology been available, it is probable that great savings might have been possible by using plastic foam backfill to "float" across the preglacial gorges with cut-and-cover construction.

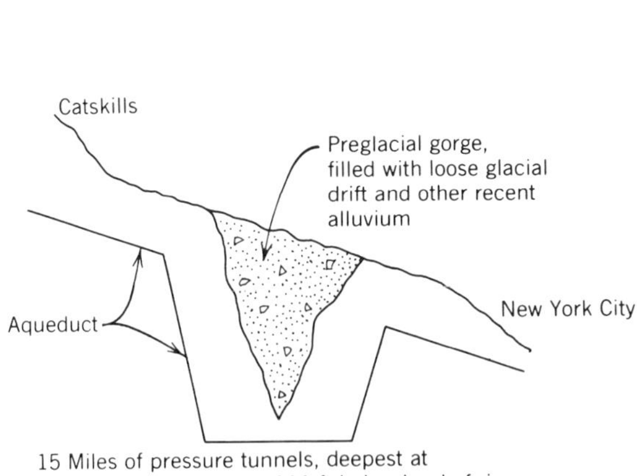

Figure 5.7. *The Catskill Aqueduct.*

Airport Construction Foam plastic subbases might be used to establish a weight credit in airport pavement design. Because many airports must be located in areas of marginal soil support—bay muds, for example—the technique may have wide applicability. (It may be noted that most stream crossings, as at the Pickford Bridge, would also involve construction of approach embankments of soft or loose sediments because of recent alluvium). For those who might consider this application, very serious attention should be given to the possibility of differential (pavement) icing. (See following section, Technical Problems—Some Suggested Solutions. See also Monahan (1993).

Retaining Structures (Figure 5.8) The use of foam plastics behind retaining structures is a logical extension of the weight-credit concept. In this case, the lateral pressure on the retaining structure caused by the pressure of the backfill itself will be reduced to a negligible value with the use of the foam plastic in place of the usual soil backfill. Thus, where no major live loads are involved, the size and strengthening of the wall required to retain the backfill will be reduced substantially. Drainage design can also be greatly simplified.

A possible application would be the use of wedge-shaped sections of precast foam plastic for various size excavations adjacent to basement walls. Or a system might be developed to fabricate or inject the foam on site. For large construction projects, the latter technique would be preferable.

Slope Stability Figure 5.9 (Wager and Holtz, 1976) illustrates a solution to a difficult slope stability problem through a unique use of reinforced earth technology. The forces P and $P \tan \phi$ providing counterclockwise resisting moments are developed by the inclusion of short sheet piles that are connected by pretensioned steel, the steel being stressed to provide a factor of safety of at

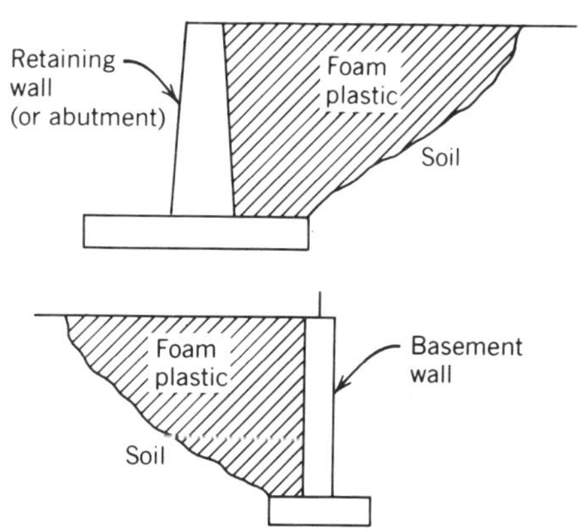

Figure 5.8. *Retaining structures.*

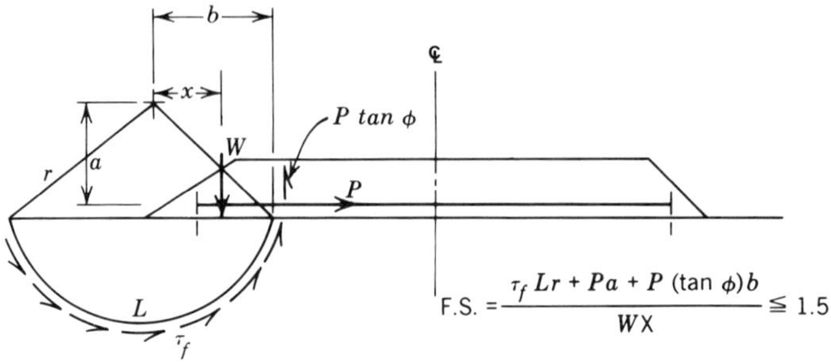

Figure 5.9. *Slope stability. (From Wager and Hotlz, 1976.)*

least 1.5. This procedure, while unique and effective, has something in common with most approaches to increasing slope stability: providing greater resistance to failure.

It is my opinion that many slope stability problems can be solved at less cost by considering the neglected part of the factor of safety, the driving forces—in this (and most) cases, the weight *W*. The use of some foam plastic in the embankment could reduce or perhaps eliminate the need for reinforcement. In cases where elimination of reinforcement can be achieved, the savings in construction labor costs could be major.

Poleset Figure 5.10 illustrates the combined use of two super lightweight fills, precast rigid boards and a material called Poleset. Both materials are foam plastics, but of significantly different composition. From my perspective as a foundation engineer, a principal difference is the method of placement. Poleset is a poured-in-place material. One of the important distinctions, as

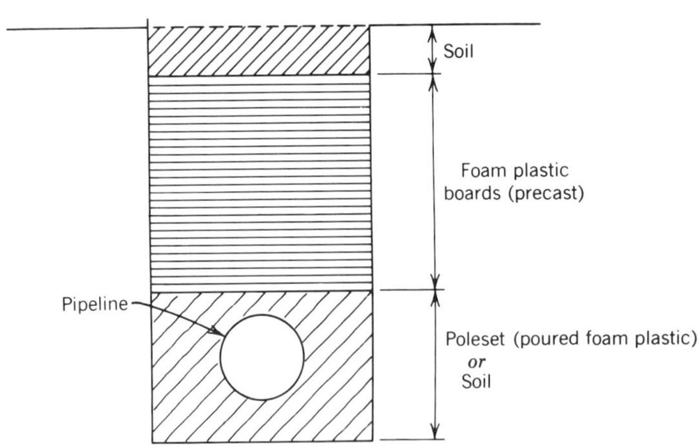

Figure 5.10. *Weight-credit versus piles. (Rospond Associates.)*

mentioned briefly in Section 5.2.1, is that Poleset would seem to be much more suitable for filling irregular and/or confined spaces. That was the initial rationale used in stipulating the poured Polset backfill under, over, and around the pipeline of Figure 5.10, with the precast boards simply stacked in place, above.

Styrofoam HI, one of the available plastic boards, is a registered trademark of the Dow Chemical Company. Some of its important properties, including cost, have been described in Section 5.2.1. Its use in weight-credit foundation construction is also described. It is covered by two patents (Monahan 1971, 1973). Figure 5.11, the Pickford Bridge, shows the stacked bundles of Styrofoam HI boards. Details of the job are described in Section 5.2.1.

More information on an assortment of rigid foam plastics, their properties, and current costs and availability may be obtained from Research Specialists, Dow Chemical U.S.A., P.O. Box 515, Granville, OH 43023.

The information I have about Poleset is sketchy, obtained initially from nothing more than a blurb of about 200 words in an article in *Better Homes and Gardens*, April 1974, p. 60. Most of what I have learned is very promising: It is very strong, very lightweight (quoted at 1½ lb/ft^3), and is reported to be waterproof. Variations in mixing ingredients can produce a wide variety of strengths, densities, and costs. This is consistent with what I had learned about foamed plastics earlier, especially the quotation from *The Modern Plastics Encyclopaedia (1968)*:

> . . . foams may be produced which have densities ranging from less than one pcf to about 70 pcf, with an almost limitless range of chemical and mechanical properties.

Figure 5.11. *Styrofoam HI approach fill, Pickford, Michigan.*

Poleset is supplied by Forward Enterprises, Inc., 9430 Telephone Road, Houston, TX 77075.

The job depicted in Figure 5.10 is for a site in Woodbridge, NJ, underlain by very soft, weak, compressible soils. Soils consultants had recommended piling for all structures, *including the utilities.* (The latter is italicized to emphasize that this is not very common practice and would be needed only for exceptionally poor soil conditions.) Having been earlier made aware of my weight-credit methods, the project engineers conferred with me, and the investigation suggested by Figure 5.10 began. This led to a computer study by Rospond Associates, which compared the costs of weight credit versus piles. Weight credit, I am informed, was cheaper by $170,000.

As it developed, Poleset was not used in Rospond's design because the boards provided enough weight credit so that normal soil fill could be used around the pipe. I do not have any cost figures yet for the Poleset, but according to Rospond, "it's expensive." With little doubt then, it is currently more expensive than the plastic boards. However, the laws of economics (higher volume, lower unit price) might tend to make Poleset more competitive in the future, particularly if its (apparently) promising mechanical and physical properties make it especially suitable for increased use as a lightweight fill in weight-credit applications.

As a result of a presentation I gave to the Geotechnical Group of the Metropolitan Section, ASCE, an application of Poleset is being considered for a job in Jersey City, NJ. The floor slab of a very old structure is virtually unusable for significant floor loadings due to the excessive settlement of an underlying loose miscellaneous fill. The tentative plan is to remove about 4 ft of the fill, judged to weigh about 90 pcf, and backfill the space with poured Poleset.

Increased interest in lightweight fills for weight-credit applications was also generated because of considerations of the multibillion dollar Westway project (a planned interstate highway section on the west side of Manhattan that would have incorporated many satellite projects in land and real estate development). This partially explains the huge cost of the federally financed project. Since the new land, about 225 acres, would have been created by filling along the eastern shore of the Hudson River (as with the earlier 93-acre Battery Park described in Section 5.2.1), the usual dredging and filling process would involve placing river muds on river muds. Because of the geologic processes involved in their formation, such muds are almost always fine-grained, soft, relatively heavy, and thus potentially very compressible. The fill, of course, would be carefully compacted under the supervision of geotechnical experts, but the soft, natural soils below would undoubtedly create concerns about settlement. The selected use of weight credit for certain pieces of the project (not major structures), might often be an alternative design possibility, an alternative to piles, for example. In fact, in 1976 I conferred with Robert Johnston, of Mueser Rutledge, Wentworth, and Johnston, New York Geotechnical Consultants, about the possibilities of using Dow's Styrofoam for just such a piece of Westway. It was he (quoted in Section 5.2.1) who said, then, "I hate to bury voids." The status of Westway, or a substitute project, remains in some doubt.

TECHNICAL PROBLEMS—SOME SUGGESTED SOLUTIONS The presently recognized technical questions relating to the use of foam plastics in weight-credit applications to foundation construction are related to permanency and durability under the influence of load and exposure to natural and man-made environments.

Special problems during service relate to differential icing, chemical resistance, long-term compression, and flotation.

Differential Icing On highway insulation installations, the purpose of the foam plastic was to prevent or minimize frost action in susceptible subgrade soils (i.e., silts). Experience has shown that the insulation can protect the subgrade, but in some instances it can cause the pavement to become icy along treated sections of the roadway, thus creating a hazardous driving condition.

Solutions could include avoidance or special designs. Also, where differential icing is considered to be a possibility, special caution signs may be used, as are customary at bridge approaches.

Avoidance It is my opinion that the extensive use of foam plastic under long sections of high-speed highways in cold regions should be avoided (or very carefully considered) because of the potential hazard of differential icing.

Dow Chemical studies suggest that differential icing will not occur in regions where the cumulative degree-days (°F) do not exceed 100. This would include most of the southern-tier states.

While each case of potential highway usage should be considered on the basis of its specific technical problems, economics, and risks (of differential icing), it is thought that the procedure would be reasonable in cold regions for short sections of low-speed roads and (especially) for grade separations (as at the Pickford Bridge) where the loadings of approach embankments can be severe with normal fills. In such cases, a sign can be posted such as CAUTION—PAVEMENT MAY BE ICY NEAR BRIDGE. Such a sign would be but a slight modification of present practice warning of possible icy pavement *at* bridges.

A case of reasonable usage for a low-speed road involved a short access road to an expanded golf course in Connecticut. Apparently because the project was an expansion of an existing facility, there was little or no choice for the location of an access road, and the area was underlain by very soft soils to considerable depths. I was advised that this road section was repeatedly surcharged (i.e., repaved) to the point where the trees on both sides became visibly tilted. The cost of such periodic maintenance was substantial and probably far exceeded what the cost of floating the road (with the use of foam plastic) would have been.

Special Designs Where circumstances dictate, special designs could include "heat bridges" to allow enough heat to leave the subgrade to prevent icing, yet provide the insulation necessary to protect the subgrade from frost action: an optimization approach. Aluminum pipes or sheets are suggested as heat bridges.

Computer studies by Dow Chemical have been developed that enable the computation of sufficient soil cover *over* the foam plastic to provide the necessary heat sink to preclude icing.

Chemical Resistance Foam plastics (e.g., Styrofoam) generally have poor resistance to some materials, most notably gasoline. As noted for the Pickford Bridge construction, polyethylene sheets were used to cover and protect the Styrofoam bundles against possible gasoline spills.

The possibilities of exposure to other materials should be considered for each case and for each type of foam plastic used. Suppliers can supply detailed information and recommendations.

Long-Term Compression A question that arises regularly in discussions with engineers who are considering the method is that of long-term compression in service—creep. Most studies have been done on the relatively thin sections of Styrofoam HI that have been used for insulation, as presented earlier, and the permanence and stability would seem favorable.

Perhaps the best (and only) evidence of stability that generates confidence for weight-credit applications (where greater thicknesses would be used) is the Pickford Bridge performance. As far as I am aware, no problems have developed.

Flotation At Pickford Bridge, 5 ft of soil fill, placed above the plastic, was sufficient pressure to pin down the foam material in the event of flooding and the accompanying rising water table. Soil pins, analogous to anchor bolts in tunnel construction, might also be employed where larger weight credits are needed.

In cases of support of buildings (as opposed to embankments, pipelines, etc.), flotation problems would not generally be acute; the design approach would be to create a positive pressure in the subsoils consistent with (their) allowable pressure tolerances. In general, zero stress increases would be avoided.

Sunlight Direct exposure of foam plastic to sunlight for extended periods of time should be avoided.

Ozone Depletion Since the first edition of this book was published, a new technical problem has arisen: the concern that chlorofluorocarbon emissions (CFCs) are depleting the ozone layer. It is estimated by some scientists that projected rates of depletion could cause major increases in skin cancers and eye cataracts, and deplete our ability to fight off infection; crop damage and disruption of the ocean food chain could also result.

One of the sources of CFCs is reported to be "the propellants that are used in the production of foam plastics." (*The Star Ledger*, Newark, NJ, 23 October 1991).

Upon learning of this problem, I called the research director at Dow Chemical, and was assured that the production methods for their product, an extruded polystyrene (XPS), was now done in such a manner as to "solve" the CFC emissions problem. However, since then, I have discovered a researcher who states, "EPS (expanded polystyrene) [different from XPS] is the only rigid plastic foam that does not use gases such as CFC or HCFC in its manufacture" (Horvath, 1992).

While these statements may not be contradictory, there certainly appears to be some question. Manufacturers and potential users are urged to resolve this apparent difference.

FUTURE USAGE It is my feeling that injected (cast-in-place) foam plastic holds great promise for future usage because of several factors. First, the materials that produce the foam can be brought to the site in appropriate containers of much smaller volume than that of the (ultimate) injected foam plastic. Second, the materials can be injected into cavities of any shape. Third, there is a certain "neatness" to the operations envisaged that suggests major savings in labor costs. Finally, such backfills will eliminate many of the problems of soil filling and compaction: texture (grain size) specifications and compaction and fill control, especially in confined spaces such as trenches and wedge-shaped fills behind retaining structures and basement walls. (It is my opinion, based on many experiences and much reading, that such backfilling procedures are very often done incorrectly and lead to many problems of a technical and litigious nature.)

On the negative side, it is thought that foam quality and quality control would not be as good as that of precast foams such as Styrofoam. Compressive strength and water absorption would have to be considered carefully and estimated conservatively. With respect to water absorption, usage might be often restricted to areas above the water table, to retain weight credits throughout the life of the structure.

Precast foams will have most usage where quality control is vital and where consequences of distress or failure are major. As with any new method, confidence in the method is necessary for extensive future usage. Usage with success will breed routine usage.

HISTORY REVISITED In the first edition, the history of the development of the use of foam plastics for foundation construction was only touched upon, largely because I was not then fully aware of activities, particularly in foreign countries. I have since learned much of what has occurred, and have concluded that it is appropriate to present an accurate and much more complete history.

In 1966, a patent was issued to G.A. Leonards, of Purdue University, for the use of foam plastic boards as insulation for protecting subgrades against damaging frost action (Leonards, 1966). The material used in subsequent applications was Styrofoam HI, a product of the Dow Chemical Company. (HI designates "highway insulation").

In 1969, unaware of Leonard's patent, I independently conceived of the idea of using foam plastic in *weight-credit* applications in foundation construction, and filed for a patent. The filing date was 12 February 1970. The patent application was initially rejected by the examiner, based largely upon the existence of Leonards' patent ("prior art"). Upon carefully examining Leonards' patent, I learned that no reference was made to weight-credit applications. Indeed, the *density of the foam was not even listed in the nomenclature of the patent.* I presented the argument to the examiner that, while the ideas were outwardly similar, they were fundamentally different. The examiner agreed, my patent was approved, and was issued on 14 December 1971. A second patent, a "continuation-in-part" (CIP), was issued to me on 24 July 1973. This patent covered weight-credit applications dealing with lateral pressures (retaining walls, etc.).

Within the next couple of years, I developed an agreement with Dow Chemical. During the same time period, Dow learned of a problem that the state of Michigan was having with a potentially unstable bridge, The Pickford Bridge, described in Section 5.2.1. Dow asked me to grant them permission to pursue this job using the methods encompassed by my patents, with immunity from infringement, arguing that a successful application would portend many future applications. I agreed to forego royalties, and granted them written permission to proceed.

As time passed, I continued to work with Dow on the development of weight-credit applications. Dow had, by this time, elevated the endeavor to "project status" within the company, and had even planned to develop a new product name. Styrofoam WC, for weight credit. Unfortunately, an incident occurred that was to bring an abrupt halt to further interest or activity on Dow's part: A major accident occurred on an icy pavement, the cause of which was contended to be the installation of Styrofoam HI under the pavement (as per Leonards' patent, which I think Dow now owned). This problem of "differential icing" is described in the previous section. Subsequently a decision was made at the executive level within Dow to abandon all "aggressive" activity using foam plastics in thermal *and* weight-credit applications. Efforts to convince people at Dow that almost all of my perceived applications had no possibilities for problems of differential icing (or could be mitigated) were unsuccessful.

This decision by Dow particularly explains why there has been so little activity in this country using my patent methods. This, I have recently learned, is not true elsewhere, and some activity has begun to emerge in the United States.

During this period (the early 1970s), the Norwegians were actively pursuing weight-credit applications using expanded polystyrene (EPS). In 1987 (the year after my first edition), an article appeared in *Geotechnical News*, the Journal of the International Society of Soil Mechanics and Foundations, published in Canada, in which a number of jobs in Norway using EPS as fill were described (Flaate, 1987). I wrote to the author requesting information, mentioning my patents, but emphasizing that I held no patents in Norway. I got no response. In January 1993, I was the lead speaker at a session on artificial fills at the annual conference of the Transportation Research Board in Washington, DC, at which time I obtained a document that was an updated description of experiences with EPS in Norway. Recently, there have been a number of articles about EPS use in several American publications, some of which cite the Norwegian experiences with EPS, starting with a job in 1972. Although this date is two years *after* I filed my first patent, some articles strongly infer that the Norwegians initiated the idea and the practice (Horvath, 1992a, 1992b; Yeh, 1992).

In addition to my patenting activity, I published and spoke about my methods starting in the early 1970s, including conference proceedings (Fang, 1976), and talks at the New York Met Section Geotechnical Group, and the New York Academy of Sciences Engineering Division. And yet, despite the fact that almost a quarter century has passed since I filed my first patent, and despite all of the intervening work, including the publication of this book in 1986, I do not know of a single publication that has cited my work!

Finally, before returning to the technical aspects of artificial fills, if you see a stocky, bald-headed guy running up and down in front of the United Engineering Center Building screaming, "I invented foam plastic as fill, I invented foam plastic as fill . . . ," you will know who he is, and why he's doing it.

RECENT DEVELOPMENTS Since the publication of the first edition (1986), there have been significant developments in both research and construction activity relating to the use of foam plastics as a weight-credit foundation material, both in the United States and throughout the world.

Research Although manufacturers of precast foam plastics have tested their products and reported their physical, mechanical, and chemical properties, there has not been a great deal of activity among geotechnical researchers to investigate the properties peculiar to the design needs of geotechnical practitioners. Recently, however, a research project was completed that establishes some important "geo-engineering" properties of one form of rigid foam plastic, expanded polystyrene (EPS).

Negussey and Jahanandish (1993) tested and compared some engineering properties of EPS to soils, comparing EPS samples of two densities to samples of soft inorganic clay and a uniform silica sand. Their results indicate that "the engineering properties of EPS can be quantified in a manner similar to earth materials." Strength and deformation behavior and lateral stress coefficients were compared.

A review of field performance, notably in Norway, confirmed much of the information regarding permanency, durability, and technical problems described in previous sections.

Case Histories Since the weight-credit jobs that I have discovered were ones that I had no direct involvement with, and thus have been reported by others, I shall supply only brief descriptions, and cite the sources for each.

EPS blocks were used to reconstruct an embankment slide that had occurred near Durango, CO. The BASF Corporation, Parsippany, NJ, supplied the polystyrene material. The job was designed and supervised by the Colorado Department of Highways (*Engineering News Record*, 27 April 1989, p. 17). Approximately 40,000 cubic yards of EPS block was used as fill for a new shopping mall in Syracuse, NY (Horvath, 1992). This article also contains a summary of current ASTM Standards relevant to EPS block.

Norwegian experience with EPS construction has encompassed more than 100 road projects since 1972 (Flaate, 1987, 1989). Canadian experience includes the use of EPS backfill behind seven newly constructed bridge and overpass abutments in soft foundation areas near Vancouver (deBoer, 1988).

A paper presented at a seminar in Oraka, Japan, describes seven case histories of a representative variety of EPS applications in Norway. Material requirements and technical advantages are described, as are aspects of durability, pavement design, and economy. The article also lists countries that have used EPS weight-credit techniques in foundation construction, with Japan being the most active ("approaching Norway in total volume of EPS used"). Other countries cited are Sweden, France, the Netherlands, the United Kingdom, Canada, the United States, the Philippines, Ireland, and West Germany (Frydenlund, 1991). (I cannot resist the temptation to add, "I wonder if any of *these* folks ever heard of Monahan?")

5.2.2 Elastizell

This material is a pumpable lightweight "concrete" produced on site by adding a liquid concentrate of hydrolized protein to a cement and water slurry. These are six classes, I–VI, with cast densities ranging from about 18 to 80 pcf. Corresponding compressive strengths are 40–700 psi.

One of the larger jobs done with Elastizell (a proprietary product) was for a bridge abutment over weak soils on I-94 near Minneapolis. About 42,000 cu yards was poured.

Typical designs incorporate more than one class, placing the stronger materials where performance requirements warrant. Elastizell does not require compaction, and, once set, it does not apply lateral pressure to walls (Elastizell Corporation, 1993).

5.2.3 Solite

Depending on the locale of its manufacture, Solite, also a proprietary product, is produced from either shale, clay, or slate. It is expanded in a rotary kiln at high temperatures to produce a lightweight, subangular granular material that is free-draining.

The material is used either as a soil fill or as an aggregate to produce lightweight concrete. As a fill, it is normally compacted to densities less than 60 pcf, yielding a material with an angle of internal friction of about 40°. The material is chemically inert. As a concrete, it s unit weight is about 116 pcf, with a 28-day compressive strength of about 6510 psi (Solite Corporation, 1993).

5.2.4 Geocell

Mearl Geocell is a "cellular concrete for geotechnical applications." It has a cast density of 36 pcf. The material is available from The Mearl Corporation, 220 Westfield Avenue, West, P.O. Box 208, Roselle Park, NJ 07204.

5.2.5 Hybrid Design

An interesting and challenging approach to design for many jobs would be to consider combinations of all of the materials described in the foregoing sections for overall weight-credit approaches. Since Solite may be used as a lightweight concrete in the main structural members of (say) a bridge, it would be feasible to design an entire project using superlightweight materials (foam plastics) and lightweight materials (Elastizell) for all fills, fitting the material selections to the weight-credit needs, and use lightweight concrete (Solite) for many of the structure components . . . a true hybrid design! Such a project would, I expect, be an exciting challenge to some of you computer buffs out there . . . Optimize, Optimize!

5.3 WASTE MATERIALS AS FILLS

As urban and industrial development occurs in a given region, good fills within reasonable haul distance become more scarce and expensive. At the same time, the very processes of urbanization and industrialization generate larger and larger quantities of what was once considered waste but is now being viewed as recycleable material. Among these products are ash, glass, rubber (tires), wood chips, aluminum and other metal containers, and a broad category one might call construction rubble.

In addition to the more common waste materials, there exist throughout the world a variety of materials of which there is an annoying abundance, materials such as sulfur, bamboo, and, of course, garbage. An interesting publication that deals with the use of such materials in construction is *New Horizons in*

Construction Materials (Fang, 1976). This publication has served as a basis for further research investigations in my work, and I recommend it for that purpose for other investigators, including students searching for interesting, challenging, and unusual thesis topics. Some will be particularly suitable for persons from the regions where the materials are abundant, for in many cases standard materials are unavailable or too expensive. In most cases, the uses of the materials may be extended to the construction of fills or their reinforcement. Earth fills reinforced by bamboo is one suggested possibility. Obviously, a certain amount of research will be necessary to bring such technology to construction applicability.

Chae and Gurdziel (1976) have studied New Jersey fly ash as a fill, and report design strengths of about 5 tsf. One of my investigations involved mixing their fly ash with natural soils of better texture to determine the engineering and compaction characteristics of the mixture. Obviously, various mixture proportions will produce fills of differing properties. Since the fly ash has normally been simply discarded (at some hauling cost), its use as a fill or fill component is a potentially attractive economic alternative.

With my encouragement and supervision, students have investigated the use of glass and rubber as fill components. Sulfur is being investigated as a replacement for asphalt in bituminous pavements, a potentially enormous cost reduction (McBee et al., 1976).

As is now well known, the use of garbage in landfills is common. Indeed, with the advent of the environmental movement, and the horrors of Love Canal, what goes into "sanitary" fills is now of major concern and study. Believe it or not, the Fritz Laboratory of Lehigh University once did a compression test on one cubic yard of compacted raw garbage (which arrived in July from California, undoubtedly to the dismay of receiving department personnel). A photograph of this "first" appears on p. 133 of *New Horizons in Construction Materials.*

A leading national magazine that regularly reports on recycling programs has stated that "current markets for most recycled materials . . . are experiencing their deepest recession in years" (*Public Works*, July 1992, p. 7). One particularly daunting problem is the continuing accumulation of waste tires, which are now estimated at 2 billion and are growing at a rate of 189 million per year (Humphrey et al., 1993). The disposal or reuse of these tires is complicated by the many potential environmental problems that are associated with such efforts, and the regulations that pertain.

Perhaps as a result of the recognition of these factors, the problems are gaining increased attention. The use of shredded rubber tires and other waste materials as fills is a very sensible approach to the problem because of the large volume of materials that can be recycled in this manner. A full session, encompassing six papers, was held at the 72nd Annual Transportation Research Board Meeting, entitled "Use of Lightweight Waste Materials for Embankments over Soft Soils," with emphases on shredded rubber tires and wood chips as fill (TRB, Washington, DC, January 1993).

Comprehensive research programs dealing with the use of shredded tires have been completed at the University of Maine, the University of Wisconsin, and Purdue University. Projects using both shredded tires and wood chips have been constructed in Oregon, Washington, and Minnesota.

The Maine research study determined the properties of gradation, specific gravity, absorption, compacted density, shear strength, compressibility, and coefficient of lateral earth pressure at rest. (The latter property was investigated because the focus of the study was for the possible use of shredded tires as retaining wall backfill.) Samples from three suppliers were studied (Humphrey et al., 1993).

The Wisconsin study, in addition to comparable determinations of engineering properties, also investigated the environmental suitability of shredded rubber. Parameters tested in lysimeter (leachate) studies included COD, BOD, CL, SO_4, pH, alkalinity, hardness, TDS, Ba, Fe, Mn, Zn, Pb, and Na. While cautioning that there is need for additional field studies, the test results "indicate that the shredded automobile tire samples show no likelihood of being a hazardous waste" (Edil and Bosscher, 1992). This study, being a general investigation of shredded tires in highway applications, also includes studies of models, test embankments, and pavement design, and lists recommendations for design and construction specifications.

The Purdue study includes compaction, compressibility, and permeability test results. An excellent comparative overview and literature review of waste and lightweight fill alternatives is followed by the observation that "information concerning stress–strain–strength behavior of tire chips for design and performance prediction of tire embankments, and long-term environmental impacts of shredded tires is severely lacking" (Ahmed and Lovell, 1993).

The Maine, Wisconsin, and Purdue studies report compacted unit weights of tire chips to be significantly lower than soil fills, with values ranging from about 20 to 40 pcf for "pure" tire chips. (The Wisconsin study also tested varying mixtures of soil and tire chips, plus layered systems.) Thus, the use of tire chips also provides weight-credit benefits for construction applications on soft soils (See Section 5.2.1).

Three recent case studies have been reported in lightweight fill applications using shredded tires and wood chips. In southern Oregon, shredded tires were used on a landslide repair project. Approximately 580,000 tires were trucked in from four sources from as far away as 275 miles. The cost of the in-place fill was $12.87 per cubic yard, but included a significant rebate from Oregon's Department of Environmental Quality. This reimbursement reflects the willingness of legislators to provide financial incentives to encourage the use of waste materials as fills, thus making such fills possibly competitive with other alternatives (Upton and Machan, 1993).

The combined use of wood fiber and geotextile reinforcement was used to construct a lightweight fill across a swamp area in Washington. Stage construction techniques, planned on the basis of careful instrumentation monitoring, were used to maintain stability during the controlled rate of construction. The wood fiber used was of the "fresh" classification, as defined by the classification criteria of the Washington State Department of Transportation. Environmental regulations required that the wood fiber not extend to a level below mean high water. Additionally, a topsoil thickness of about 2 ft was placed over the wood fiber to guard against fire and exposure to oxygen. Despite heavy logging traffic, no serious pavement distress has been noted in the approximately 5-year service period. Water sampling of leachate, and physical inspection of the wood fiber near the surface indicate no pollution or wood chip deterioration of consequence. A cost saving of approximately $500,000 was realized over the net lowest viable option. The writers assert that "wood fibre can be used for permanent applications with design lives in excess of 50 years" (Allen and Killian, 1993).

In Minnesota, a job was completed that used wood fiber, shredded rubber tires, and geotextiles in an embankment design that was constructed to cross weak peat soils. Geotextile was placed at the bottom of a 5-ft excavation, and wood chips were placed to a height of 1 ft above the water table, as required by the Minnesota Pollution Control Agency. Shredded tires were then placed to a height 3 ft above the original road surface. The tire layer was covered with geotextile, and the fabric was sewn together with the lower fabric to form an enclosing "bag". The project was designed and supervised by Toltz, King, Duvall, Anderson and Associates, Engineers Architects Planners, of St. Paul, MN (Monahan, 1993).

5.3.1 Low-Level Geothermal Heating/Cooling

The use of fills for energy conservation in "second homes" is a subject that has interested me for some time. Many vacation homes in temperate regions of the United States are empty for a substantial portion of colder time periods, but are kept heated to protect against structural damage such as burst pipes during cold weather. Typically, many such homes are kept at 40° F during these periods of absence. During the summer, when occupancy is most frequent, the need for air conditioning is common.

The cost of this energy must be substantial when one considers the number of such homes in the United States, and that portion that goes toward "absentee heating" is particularly wasteful of money and energy. I believe there is a much cheaper and energy-conserving alternative: tapping the "free" energy a few foot below. As all geotechnical engineers are aware, the temperature of the ground a scant several feet below grade is almost a constant 40°F all year-round, i.e., below the frost line. It remains to develop a scheme to take advantage of this free geothermal energy. I believe this can be done by a combination of inexpensive construction techniques using materials of differing thermal conductivity, some of which could be waste fills.

Figure 5.12 shows a schematic of a small structure that represents a second home, for simplicity showing just one interior wall. The simple scheme, as shown is to use materials of very high thermal conductivity in the excavated

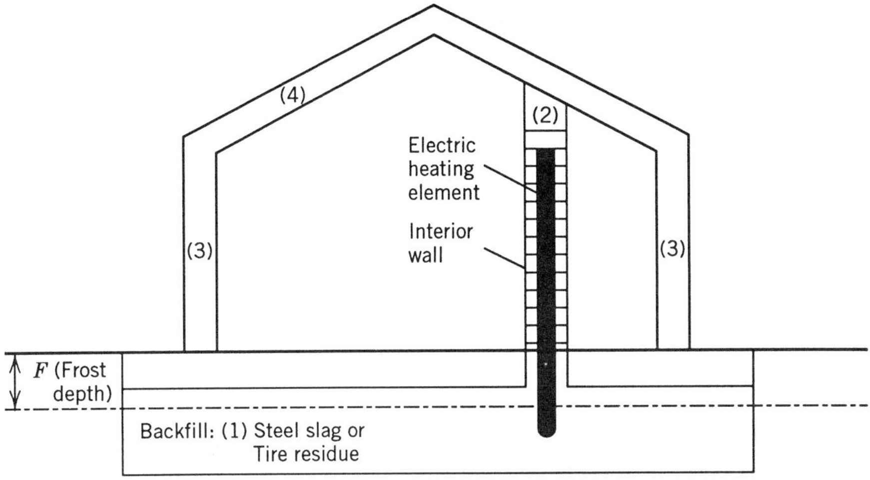

Figure 5.12. *Low-level geothermal heating and cooling.*

space beneath the frost line, with other material of high thermal conductivity extending upward into the interior walls of the structure. The roof and all exterior walls would, of course, contain materials of very low conductivity, i.e., insulation. For backup protection, the interior walls could be instrumented with some type of heating elements, much like the common dipstick heater for automobiles.

Possibilities for the different materials that have occurred to me are the following. (Numbers refer to the spaces in the figure.)

(1) *Steel slag or "tire residue."* The latter is a name that I have coined for a material that was described in *Science News*, 25 April 1992, p. 259, in response to a letter to the Editor. The writer had inquired about the process of separating steel from waste tires, and was told that the "Oxford Energy Company in Dearborn, Michigan burns the tire with steel belt and bead, then collects the metal's residue ash from the furnace and sells it as a by-product." Another process, according to BAS Engineering Consultants, Inc., involves the separation of the steel from preshredded tires by passing the chips through a magnetic field, which separates the metal from the rubber. I have not determined what the costs or availability of these waste materials are, or what the specific thermal and other properties are, but they are both clearly metallic, and thus likely to have high thermal conductivity.

(2) *Metallic sheet material, or perhaps ceramic*, both of which would have high thermal conductivity.

(3) *Stacked telephone books, or "rubber plastic."* The former may not be as unusual as it might first strike the reader, inasmuch as telephone books have already been used for this purpose in Taiwan, by way of Tucson, AZ, according

to an article in *The Pocono Record* entitled, "Old Telephone Books Don't Die." The city collected 180,000 books, sold them to a salvage company for $4500, which then shipped them to Taiwan. I do not know what the *R* value of stacked phone books would be, but my intuition suggests that it is very high. (While on the subject of unusual ideas, I thought I'd share with you the following letter, addressed to Marilyn vos Savant, which she put in her column of "favorite letters": "I've heard that because rabbits have a high body temperature and their body heat dissipates very fast through their long ears, they've been used to heat houses. Can you tell me how many rabbits it would take to heat a three-room apartment when the temperature outside is 45 degrees? F.K., Columbia, MD.)

The "rubber plastic" is an original idea that envisions the development of a molded plastic with tire chips as a matrix. Since both materials are well known for their insulating properties, the combined material should also have excellent insulating properties. An added benefit is that it constitutes a new use for waste tires. Perhaps some plastic molder will find this idea intriguing.

(4) *Normal insulation of R value comparable to the exterior wall insulation.* This suggestion recognizes the fact that the placement of phone books in a pitched roof may be problematic. Possibly, the "rubber plastic" could be used effectively, also.

(5) The heating elements shown in the figure represent a "backup" system that might be needed during periods of extreme cold weather. It would be most sensibly used in a "stored heating" fashion, wherein the heat would be turned on, perhaps remotely or even automatically, for short periods during the night when energy costs are lower. The heat generated would be designed to keep the unoccupied house at or near the "safe" temperature level of 40°F throughout the day. This concept of "storage heating" is a technique commonly espoused by power companies (PP&L, January 1988).

Other features of design would be the use of the highest quality insulated windows and doors. Opaque, heavy, indoor shades, pulled down during non-occupancy, are also useful devices to minimize heat loss, in that "dead air spaces" are created between the windows and the shades, which serve as excellent additional insulation. Lastly, a feature that should be considered (and is often overlooked) is the insulation of exterior electrical outlets. A product I discovered recently that claims to solve this problem effectively is called a Perma-Flex Care Cover. It is a plate that covers the receptacle holes with a spring-loaded trap that cuts off the air flow from outside. (According to the manufacturer, "one study concluded that outlets are responsible for 20 percent of a home's air infiltration—almost twice what comes in around the windows.") The device is patented, and won an innovation award from the federal Department of Energy. An additional benefit is cited by the manufacturer: It protects children against electrical shock. Cost: about $3. Address: K-Products Group, 724 Commerce Street, Aberdeen, SD 57401. (*Note:* A much cheaper alternative is the use of plastic "dummy plugs." These are available at K-Mart and other stores for a few cents apiece. However the safety feature of the Care Covers may be worth the extra cost to those with toddlers.)

A few data may be useful to illustrate the potential of this scheme.

I instrumented a 4-ft-high crawl space for several weeks in the summer of 1977. The outdoor temperature range for the period was 46–85°F (39°). Simultaneous readings near the ceiling of the crawl space yielded a range of 61–66°F (5°). This almost constant temperature was maintained in a crawl space that had no special construction features such as described for Figure 5.12. Indeed, the excavation for the construction did not extend nearly deep enough to reach a level beneath the frost line, nor were there any special fills or heat transfer devices installed. The floor of the crawl space is poured concrete. Its area is about 600 sq ft.

According to *Harper's Index*, 5 August 1991, 13% of the electricity consumed in the United States is for air conditioning. Thus, the combined savings for both heating and cooling are thought to be substantial over the life of the structure.

One of the things that has been somewhat puzzling to me is the virtual absence of any attention to the analysis and design of fills for energy conservation by geotechnical engineers and researchers. What little work I have seen is usually in architectural publications, and usually involves some one-of-a-kind, avant-garde structure of considerable aesthetic, visual impact, one that just happens to also conserve energy . . . commonly referred to as an "earth-protected" structure. What I am espousing here is simple common sense energy conservation applied to very common houses, of which there are probably many thousands that would benefit. And I think geotechnical people could develop the rigorous scientific analysis that would result in major energy savings across the United States, especially in northern tier states. One way to approach the problem would be to envision the house in Figure 5.2 as simply an upward projection of the earth, rather than a structure, sort of a cross section of a ridge. Then one could observe that the soil, of thickness F (depth to frost line), has an R value sufficient to insulate the region below to maintain a temperature of 40°F.

The next question: What is that R value? Once determined by some bright young scholar out there, the trick would be simply to choose materials of comparable R values to encompass the exterior of the structure, with suitable materials of high thermal conductivity to ensure the result, throwing in a few "dipsticks" to augment Mother Nature's efforts. Thus, in such a scheme, we would not only save significant amounts of energy, but we would also utilize materials that would otherwise present problems of disposal . . . two birds with one stone, so to speak.

Finally, I predict that, unless there is a major breakthrough in energy technology, such as the development of fusion, we will again see the situation that prevailed in the mid 1970s during the OPEC crisis, when people waved at each other in long gas lines, sometimes even with all five fingers. The reason I am somewhat pessimistic in this regard is that we seem to have gone back to our profligate ways with our energy consumption. The sight of a 40-ft white stretch limo idling at the curb in the incredibly congested Manhattan Theatre District is, in my view, a preposterous absurdity that illustrates well this profligacy. If we can force a lot less waste and encourage a little more conservation, perhaps Desert Storm II will never materialize.

I am not expert in heat transfer theory and design, but my intuition tells me that the scheme depicted herein will work. It has been my intention for some time to pursue a field demonstration to prove that it works. Perhaps some energy guru in Washington will read this and regard it as a research proposal.

5.4 EFFECTS OF MECHANICAL LABORATORY COMPACTORS

Anyone who has ever done a modified Proctor compaction test alone and by manual means will appreciate the assertion that it is hard, boring work. Assuming five points would be needed to define the compaction curve (i.e., optimum moisture content), the 10-lb hammer would have to be lifted 18 in. 625 times. Because of the hard physical work and the tedium, there are opportunities for error from several sources: not lifting the hammer the full 18 in. (especially as fatigue sets in); not executing a vertical free-fall, and thus reducing the energy of impact through side friction between the handle and sleeve of the hammer; slinging the hammer (especially if the radio is on to lively rock music!); losing count (blows or whole layers); and differences in coverage patterns among individuals.

Apparently recognizing a need, equipment manufacturers have developed a mechanical compactor that can be set to deliver automatically a given number of blows from a given height, with provision made for interrupting the action for the addition of successive layers. While it is apparent that human errors are rectified by using the mechanical compactor, ASTM procedures allow for either manual or mechanical means of compaction. It is my belief that this optional feature of the test procedure is unfortunate, in that major differences in results can occur because of variations among individuals opting for the manual technique.

I am not aware of any studies to determine differences between hand compaction and machine compaction. Such in investigation would require an extensive amount of work, for we are here dealing with human variances. Many tests would be needed to establish statistically valid conclusions. Also, we have a catch-22 situation: If ASTM was to eliminate the option by requiring machine compaction (a step that I recommend), a study would be needed to correlate future test results (by machine) with past results (mostly manual). The alternative would be to simply regard all past results as worthless.

Finally, another unfortunate feature of present ASTM procedures is that the tester does not even have to stipulate (on the data sheet) which option was used for compaction, thus rendering the work worthless because of uncertainty on the part of the potential user.

5.5 DENSITY GRADIENTS

When a soil is compacted in the laboratory (or in the field), the lower layers (lifts) will be compacted to a higher density because the zone of influence of the hammer (roller) exceeds the layer (lift) thickness. This being the case, it would seem reasonable to pass a field fill even if the field density test value is less than the target value, as long as the test value is within an established range, as determined by a carefully conducted research investigation of density gradients.

5.6 GEOSTICK CORRELATIONS

The Acker Drill Company makes a device called a geostick, which is a combination penetrometer, geologist's pick, and field sampler (for soft soils). The penetrometer feature allows one merely to push the conical tip vertically into a soil either under the weight of the stick for soft soils, or with the weight of the operator for stiff or dense soils. Based upon the penetration values, a presumptive bearing capacity is read (or usually interpolated) from a table printed on the barrel of the device.

I have begun very preliminary studies to attempt to correlate geostick readings with compacted densities, and to investigate density gradients.

Since the zone of influence, or pressure bulb, for the conical tip is at most a couple of inches, such readings should never be considered a complete substitute for direct means of field-density testing. Also, the proximity of the rigid wall of the laboratory mold may significantly complicate correlations between laboratory and field readings. However, I feel that the technique could serve as a valuable additional aid in fill control work, as long as reasonably valid correlations are developed and good judgment is used in recognizing its limitations and uses. For example, one would be foolish to arrive at a site and pass a 10-ft-thick embankment based on a geostick reading at the surface. If, however, one has *observed* the compaction process, lift by lift, and time is of the essence, the geostick reading, if previously correlated, might be a valid substitute.

5.7 DENSITIES OF UNCOMPACTED FILLS

In succeeding chapters, I will describe some problems associated with compaction in confined places, such as in trenches and behind retaining walls, where "lunchtime fills" can be a vexing field problem. These are situations where large thicknesses of fill can "materialize" while the inspector is out having a Big Mac. (If said inspector tests the surface of this fill, and passes it on that basis, I have some property at Okeefinokee Estates I'd like to sell him!)

In Section 6.5.2, I describe a case study where soil was dumped into a trench without any compaction, and the forensic challenge was to estimate the potential consequences. In such a case, all too common, it is almost always necessary to assume an initial density condition. In the case described, a relative density of 40% was assumed.

In a more recent case, a large sewer job (Case Study 4, Section 4.2.4) had been done that required a very deep trench excavation, resulting in a surface backfill area that was of considerable areal extent. Specifications for the manner of backfilling were adequate, but, as is also all too unfortunately the case, there was inadequate provision for on-site inspection of the work. Investigations for the housing developments that were to be constructed over the area led to concerns about the condition of the underlying backfill, and experts were brought in to evaluate the situation. (I was one.)

As described earlier, extensive amounts of time and money were spent by a number of participants in evaluating existing records of construction, and, in fact, a rather complete and sophisticated site investigation was done for the sole purpose of determining whether the sewer backfill was done according to specifications. A nationally renowned geotechnical engineer supervised this study, and concluded that the backfill was "very loose, apparently uncompacted." He recommended site remediation with the use of heavy vibratory rollers, so as to compact the soil to a sufficient depth to permit the housing construction with assurance of future stability. As with the case of the simpler trench backfill, the major focus was essentially a determination of the density of uncompacted fills. As noted, for small jobs, this can be merely a matter of inspecting and evaluating existing records, but for large jobs, the process can become extensive and expensive, as for Case Study 4.

As far as I am aware, no work has been done to establish what can bluntly be described as "dumped densities"; or, to put it a little more elegantly, there is no research project entitled "A Study of Field Index Properties Defining Uncompacted Fills." I think this would be a relatively simple and inexpensive project, and in this imperfect world in which we live, where "lunchtime fills" do happen, it would be a worthwhile endeavor. The approach to the project could be to intentionally dump fills, and then test them in accordance with the commonly used field tests. These could include standard penetration tests (SPT), static cone penetrometers (CPT), and a new and promising method of field testing, dilatometer test soundings (DMT). The latter two field testing methods were used in Case Study 4. (Schmertmann and Crapps, 1986). As presented in Sections 4.4.1 and 4.4.2, the blow counts N, (from SPTs) were estimated to be about 27 and 50 (+) for SP and MP, respectively. Thus, anything significantly less would infer inadequate compaction. However, these data are skimpy, and, while useful, would not be nearly so worthwhile as specific numbers for "dumped" fills, as would undoubtedly be obtained through a project such as I suggest.

Actually, the results could also be applicable to another area of geotechnical construction, where loose fills are common and unavoidable: hydraulic fills (see Section 6.4.1 for a brief description). Liquefaction studies could be an offshoot investigation of this project. (While working on Case Study 4, I was advised that someone in Canada had done studies of the density conditions of hydraulic fills placed underwater. Such studies would be very interesting, inasmuch as this would likely represent the very lowest dumped density condition, since the effective unit weight of the soil would be about halved by submergence. Unfortunately, I never did learn of the specifics of the work, or where it may or may not have been published. However, I thought it worth mentioning here, so that someone who reads this, and is interested, might be able to locate the source of the work.)

5.8 PERCENT COMPACTION SPECIFICATIONS FOR CLAY FILLS

In Chapter 4, I presented a method for specifying percentage soil compaction for *structural* fills, which by definition are granular or essentially granular soils (EGS). The method is not applicable to clays, essentially cohesive soils (ECS), and its use is questionable for a gemicoss. It is natural, therefore, to think in terms of developing a distinct (perhaps analogous) method for application to compaction of cohesive soils.

In response to my suggestion, and under my supervision, Zwingle (1981) has proposed such a method. Because the work is preliminary in nature—Zwingle himself calls it a conceptual methodology—I will not attempt to describe his work in detail. I will, however, provide some of its highlights and provide some commentary and suggestions of my own.

1. Analogous to relating percent compaction for structural fills to density, the percent compaction for clays is related to the unconfined compression strength. For rapid loading, the unconfined compression strength is commonly taken as equal to the allowable bearing capacity for clays.
2. A mold of modified dimensions is used to permit valid unconfined compression tests on extruded, compacted samples. This is necessary because of a length-diameter ratio requirement not met by the original Proctor mold.
3. The liquid limit is postulated as a basis for correlating the required unconfined compression strengths with water contents and, in turn, with compacted dry densities of modified Proctor curves.
4. The important differences in engineering properties of clays compacted wet and dry of optimum, at *equal* densities, is considered. These behavioral differences are largely attributed by most authorities to differences in particle orientation, with the wet side exhibiting a more parallel orientation than the dry side. As a result, clays compacted on the wet side are more highly anisotropic, particularly with respect to shear

strength and drainage. Clays compacted to high consistencies (stiff, hard) on the dry side may exhibit detrimental expansiveness when subsequently wetted.

As may be inferred by the foregoing, the compaction characteristics of clays is inherently much more complex than that of essentially granular soils. I would, therefore, recommend that geotechnical specialists be consulted when dealing with such soils. I would especially caution nonspecialists to refrain from making judgments on projects dealing with clay fills where the consequences of failure are severe, notably with earth dams.

5.9 SUMMARY

In this chapter, I have presented descriptions of work that I have been involved with over about 34 years. The range of completeness extends from published work and patents to suggestions and ideas for further work. References on artificial fills are cited within the chapter. Some other references and data have not been published, but I offer to share them with serious investigators who wish to pursue the work further.

5.10 GLOSSARY

Anchor bolts Bolts that are used to pin potentially weak soil or rock to an area (below or within) that is judged to be stable; the bolts are placed or driven into the stable area and there anchored with injected grout (thin cement) or special epoxies (a more recent technique).

Bentonite A clay of very high plasticity of the mineral family montmorillonite; plasticity index is about 400. In construction parlance, it is called "drilling mud," or slurry, and is used often for preventing the walls of excavations (drill holes, trenches) from collapsing. Mixed with water (in large amounts because of such high plasticity), it forms a viscous fluid that exerts lateral pressure on vertical walls.

Excess porewater pressure When water is confined in the small pores of a relatively impervious soil such as clay, and an external load is applied, the stress induced by the external load is initially "accepted" by the water in the pores. If one were to measure the water pressure at a point, say, 20 ft below the existing water table, the pressure would be in excess of hydrostatic (i.e., greater than 20 γ_w). The amount in excess of *hydrostatic* pressure is termed the excess pore water pressure. As the water drains, or is "squeezed out," of the soil–water system, under the influence of the excess porewater pressure, the pressure would diminish, eventually reaching that of the *hydrostatic* condition. The process of water being squeezed out of the system in such a manner is called consolidation. (See Section 6.2.6 for further explanation.)

Preconsolidation (sand drains, wicks) A foundation treatment for inducing consolidation before construction (hence, preconsolidation), as described above. The external load is a surcharge, usually a fill at the surface, and the sand drains or wicks are designed and installed to accelerate the process to allow construction to commence after a reasonable time period (usually a year). See also Section 10.1.

Structural slab A concrete floor that is designed with enough reinforcing steel to span supporting columns without distress or unacceptable deflection. Slab-on-grade construction means the floor requires support of a fill. *Structural* beams and *grade* beams have the same connotation.

Uplift pressure When water flows under a structure such as a dam, upward pressures develop. These uplift pressures must be considered in design.

(+200)(−200) This nomenclature refers to sieve numbers. The number on a sieve (as 200) refers to the number of openings per square centimeter. Thus, the larger the number, the finer is the sieve. The 200 sieve is the approximate separator between fine sand and coarse silt. Thus, anything passing the 200 sieve (−200) is fines—either silt or clay or a mixture of both.

CHAPTER 6

Fills and Fill Compaction

As described in Chapter 3, there are a number of categories of natural soils that are used as fills, with differentiations based principally on texture (grain size distribution for cohesionless soils, and plasticity (PI) for cohesive soils). Of increasing importance are fills composed of waste products and artificial materials (Chapter 5). Each fill type requires different methods of field operation.

6.1 STRENGTH, STABILITY, AND IMPERVIOUSNESS: CONTRASTING REQUIREMENTS

A helpful way to consider the contrasts among fill types is to consider the components and functions of an earth dam (Figure 6.1). There are two requirements of the dam: It must be strong enough to impound the water in the reservoir, and it must do so without unacceptable seepage losses.

The most permanently stable soils are those that have a granular (cohesionless) texture, are dense, and free-draining. Marginally draining soils, such as fine sands and silts, are susceptible to instability associated with seepage and capillarity. Clay soils, cohesive and plastic in nature, exhibit potentially damaging changes. When these soils are soft, consolidation can occur, leading to perhaps intolerable settlements. When they are hard or overcompacted, the same soils may expand when wetted.

The simplest conception of the design and construction of an earth dam is to provide strength and stability using granular soil for the shells and ensure imperviousness with the clay core wall (see Figure 6.1). A cutoff wall (sheet pile or grout curtain, typically) may be necessary to reduce seepage under the dam to an acceptable level.

I further suggest that the suitability of fills—indeed, all soils—should be considered on the basis of function rather than whether they are good or bad. Clearly, clay is a good fill in the core wall of an earth dam, but bad as a backfill behind a retaining wall.

6.2 POTENTIAL PROBLEMS WITH EARTH STRUCTURES

A second reason for focusing on fills in earth dams in this chapter is the current National Dam Inspection Program, initiated principally because of major dam failures. It is probable that this inspection (and remediation) program will require decades of attention. It is also sensible that the construction of new earth dams should be done with careful attention to proper methods of compaction and familiarity with the consequences of poor design and construction will be helpful. To some degree, the material presented in the following sections will also be helpful to those involved in the ongoing inspection of existing dams.

The historic floods of 1993 are in process as I write, and the full extent of the damage is yet to be determined. Included among the damage are the failures of many levees, earth structures that are constructed along rivers to protect adjacent lands during floods. Many of these levees will have to be repaired or virtually rebuilt. Thus proper methods of fill construction take on an even more urgency than before.

6.2.1 Marginally Permeable Soils

Marginally permeable soils include fine sands and silts, with permeability coefficients in the 10^{-3}- to 10^{-4}-cm/s range. The problems associated with such soils are potential quick conditions (*boiling*), liquefaction, and frost heave. With respect to earth dams, the problem of boiling is paramount, because it is the start of an insidious process of internal deterioration of the dam. If not detected, boiling may progress to piping, caving, roofing, and eventual breaching (a sudden, complete collapse).

THE FLOW NET Figure 6.2 shows a *flow net*, representing the flow under a sheet pile caused by the impoundment of 12 ft of water. The flow net is composed of intersecting sets of *flow lines* and *equipotential lines.* A flow line is straightforwardly defined as any path traversed by water flowing from headwater to tailwater, thus *ABCD* and *EFGHIJK* in the figure; *LM* is also a flow line, a boundary flow line (as is *ABCD*). Lines 1–2, 3–4, 5–6, 7–8, 9–10, 11–12, and 13–14 are equipotential lines and may be defined graphically as demonstrated by the piezometers shown in the figure: At any point on line 3–4 (*N* and *Q*, shown) the energy of the water may be represented by its position (Z) with respect to the reference, and its pressure head (p) as represented by its rise in the piezometer tube. Thus, a particle of water at *E* will have lost 2 ft of head in flowing

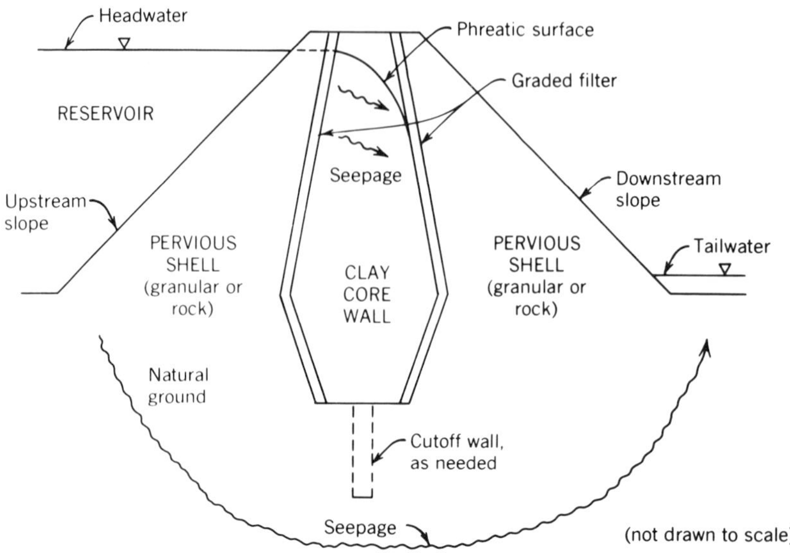

Figure 6.1. *A typical earth dam.*

from point *E* to point *F*. This would represent an *equipotential drop* (ep drop). The same particle of water, traversing the flow line *EFGHIJK* would suffer six ep drops, losing 2 ft of head across each drop (by frictional resistance to flow), for a total head loss of 12 ft (to the tailwater). With respect to fluid mechanics, note that the velocity head, $v^2/2g$, is not a factor in flow through soils, inasmuch as the velocity is negligibly small, laminar, and governed by Darcy's law.

The principal requirements of any flow net are (1) that all flow lines and equipotential lines intersect at right angles, and (2) that all elements are square. (Actually, this is true only for isotropic soils—and these do not exist. However, through a relatively simple process of "transformed sections," the correct flow net can be obtained. This is a procedure whereby the cross section of the soil profile is redrawn on the basis of the ratio of the estimated maximum and minimum permeabilities of the real anisotropic soil. The directions of these permeabilities are orthogonal, typically horizontal and vertical, respectively, in sedimentary soils. After the flow net is drawn on the transformed section, the "real" flow net is drawn by plotting all intersections of flow lines and equipotential lines back onto the true dimensions of the profile. The flow net thus obtained will not exhibit orthogonal intersections or square elements.) An element is defined as being bounded on two opposite sides by flow lines and equipotential lines (e.g., *FG* 46, Figure 6.2). Since all lines in the flow net are curved, a better definition of square is that a circle may be inscribed in each element of the flow net, as shown.

A *flow path* is defined as the region between two adjacent flow lines (*AE*, *EP*, and *PM*, Figure 6.2). Thus, Figure 6.2 exhibits *six* equipotential drops and

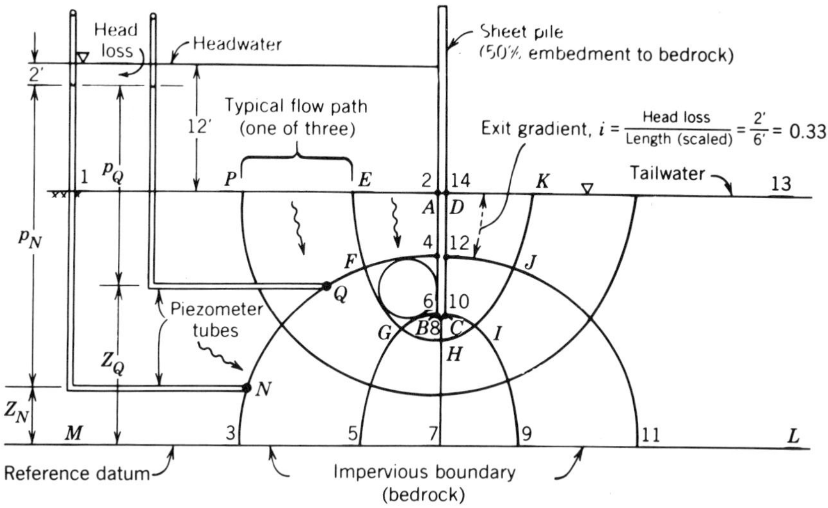

Figure 6.2. *A flow net.*

three flow paths. Partial ep drops and/or partial flow paths are valid, but in such a case all elements within the "partial" must be similar rectangles rather than squares.

The ratio of the number of flow paths N_f to the number of ep drops N_e is termed the *shape factor* for the flow net. One of the questions that commonly occurs to those learning to draw flow nets is, "How many flow lines (paths) should I draw?" There is no correct answer, as it depends on the particular case, and experience is needed to decide. However, I suggest that most flow nets should contain between two and perhaps as many as seven flow paths. It has been my observation that most beginners attempt too many flow paths for a given problem. Irrespective of the number of flow lines selected, however, you should note that there is only one correct shape factor. Thus, in Figure 6.2, if one chose to draw six flow paths, this would, of necessity, result in 12 ep drops. The shape factor would be ½ in either case.

An assist to the beginner in drawing flow nets is to study flow nets correctly drawn by others. There are a surprising number of rather badly drawn flow nets appearing in published books. See how many you can spot! *Remember:* ep's and flow lines must be orthogonal *everywhere*, and all elements must be square (or similar rectangles in partials).

A feature of seepage analysis and design that is not generally known is that the seepage *quantity* one computes from a flow net is rather insensitive to the quality of the flow net. Thus, one can get a pretty good answer for discharge even with a quickly sketched, poor flow net. However, to get acceptable answers for seepage *pressures*, the flow net must be correct, or at least more nearly so than the usual 15-min sketched variety. This information is valuable in that much time can be wasted on producing a correct flow net, only to discover that the discharge is unacceptably high. I recommend that one sketch a quick flow net and check for discharge. When the proposed design is acceptable from that standpoint, the flow net can then be redrawn carefully for purposes of pressure and stability analyses. For example, in Figure 6.2, if the discharge had proven to be unacceptably high for the 50% cutoff depth shown, a second trial at, say, 75% cutoff could be tried, for which, of course, there would result a different flow net. This approach could save days of wasted effort, since drawing a good flow net can often take many hours, particularly for a beginner.

Seepage Loss The seepage loss (discharge) is given by

$$Q = k \frac{N_f}{N_e} HL$$

where k is the permeability coefficient (velocity units), N_f is the number of flow paths, N_e is the number of ep drops, H is the total head loss, headwater to tailwater (12 ft in Figure 6.2), and L is the appropriate dimension perpendicular to the plane of the flow net (e.g., length of dam or sheet pile).

Seepage Pressures When water seeps through a soil, a seepage force develops in the same direction as the flow. The unit seepage force or seepage pressure is

$$j = i\gamma_w$$

where i is the hydraulic gradient and γ_w is the unit weight of water (62.4 lb/ft^3 or 1 g/cm^3). The hydraulic gradient is defined as *the head loss divided by the length over which it is lost.* Since seepage studies involve the determination of hydraulic gradients at points anywhere within the flow, as defined by the flow net element at the region of the point, I have found that people make fewer errors if they remember the word definition of the hydraulic gradient rather than reducing the definition to a formula. Thus, the hydraulic gradient at the center of element *FG* 46, Figure 6.2, is 2/6 = 0.33 (i.e., 2-ft head loss, divided by 6 ft, the length over which it is lost) (as scaled from the flow net element).

Of particular importance in seepage studies is the *exit hydraulic gradient*, as compared to the *critical exit hydraulic gradient* i_c. The exit hydraulic gradient is the one associated with the smallest square at exit (tailwater), since this would represent the maximum exit gradient (smallest length over which it is lost). This gradient in Figure 6.2 is $^2/_6$ = 0.33. The critical exit hydraulic gradient is determined by considering the force balance at exit. The seepage force j would be upward, corresponding to the direction of flow, and the resisting downward force would be the effective weight of the (submerged) soil γ'. The critical exit hydraulic gradient would occur when the seepage force equals the effective unit weight, or

$$i_c \gamma_w = \gamma'$$

Thus,

$$i_c = \frac{\gamma'}{\gamma_w}$$

Since the submerged, effective unit weight of soils would be approximately 60 lb/ft^3 (in accordance with Archimede's principle), the critical exit hydraulic gradient is approximately 1. If such a condition exists through poor design (or no design at all in the case of many old dams), boiling will develop. (In modern seepage design, the facility is designed so that the exit hydraulic gradient does not exceed 0.8, thus incorporating some measure of a factor of safety.) Boiling is the phenomenon that is misleadingly called quicksand in old Tarzan movies, and in many lay publications. It should be apparent from the foregoing, however, that the instability is a condition rather than a material. If the condition is not permitted to develop, the same "quicksand" can be quite stable as a foundation material. The general approach to preventing this condition (i.e., reducing the exit gradient to 0.8 or less) is to force the water to flow through a longer distance by, for example, installing or deepening a cutoff wall.

In the initial stages of boiling, individual particles of sand will be dislodged upward by the seepage forces, and be deposited in a more-or-less symmetrical pile that very much resembles an ant hill. Allowed to continue, the next stage of deterioration will be *piping*, the backward progression of a small tunnel (pipe), starting at the tailwater and progressing backward to the area under the structure (see Figure 6.3). As the piping becomes more pronounced, the flow will concentrate in the small area of the pipe, and, in accordance with the continuity principle of steady-state flow, the velocity will increase proportionally ($V_1A_1 = V_2A_2$). Since V is proportional to i, the hydraulic gradient, and $j = i\gamma_w$, the seepage forces will increase correspondingly. It may be said of piping, "The worse it gets, the worse it gets." In time, the piping will migrate under the structure, grow in size, and eventually *caving* and *roofing* (roof collapse) will occur. Finally, the structure above will settle. In the case of a typical earth dam, Figure 6.1, the clay core wall will develop cracks, and piping will commence *through* the dam, possibly resulting in breaching and complete collapse.

The kind of deterioration and distress just described is judged by experts to have been the cause of the Taccoa dam failure.

Summary The foregoing brief descriptions of seepage are included principally to emphasize how important the proper compaction of earth structures, especially dams, can be. For a fuller treatment of all aspects of the subject, Cedergren's *Seepaage, Drainage, and Flow Nets* is highly recommended, particularly for its practical approach to design and construction.

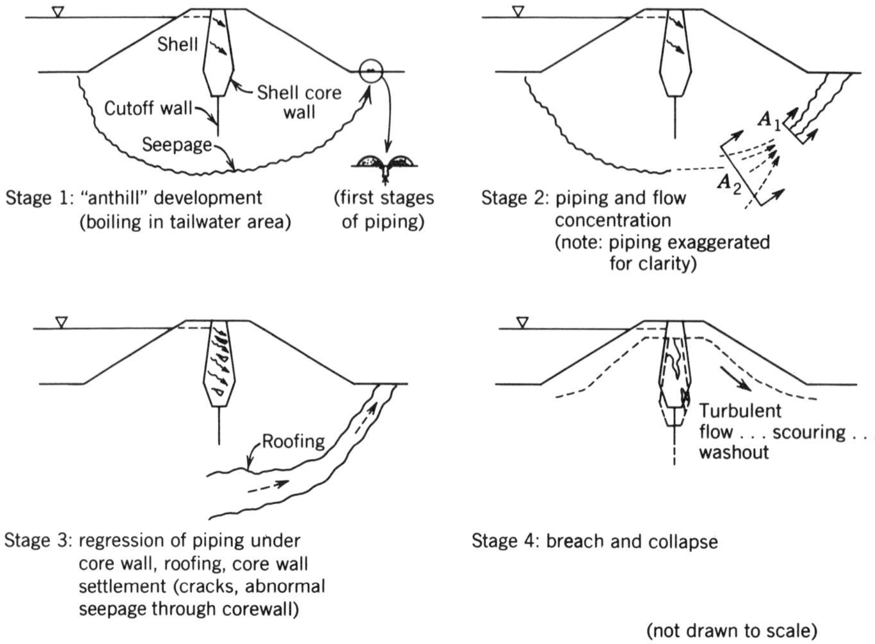

Figure 6.3. *Deterioration of an earth structure.*

LIQUEFACTION If an earth mass of marginally permeable soil is loose, saturated, and cohesionless, and is subjected to a shock, the soil mass may instantaneously lose its shear strength and fail suddenly (and often dramatically), a phenomenon called *liquefaction*.

Shocks can be caused by earthquakes, the most dramatic example of which occurred in the 1964 quake in Nigata, Japan, where 6-story buildings sank into the ground or were turned over. An aerial photograph of this dramatic event is shown in Figure 6.4. However, liquefaction can also be induced by less exotic sources of shock, involving everyday, routine construction operations. An excellent example is described by Sowers (1979), an incident involving the abrupt "swallowing" of a bulldozer traversing a loose, saturated hydraulic landfill. (Contractors and dozer operators please note!) Other common sources of shock include blasting and pile driving.

FROST ACTION Three factors are necessary for frost action to occur: a frost-susceptible soil, a shallow water table, and (seasonally) cold weather. The phenomenon is caused by the growth of ice lenses that are fed by upward capillary flow from the water table during periods of cold weather. The combined effect of the ice lenses is to produce frost heave, which is potentially damaging to any structure in or above the soil. The worst damage, however, usually occurs in the early spring. Since thawing will occur initially near the surface, the meltwater will be temporarily trapped by the still-frozen subsoil.

Figure 6.4. *Liquefaction: Nigata earthquake.*

During this time period, severe damage can occur to highways by an action called *pumping*, the high-energy ejection of water and soil in a lateral direction caused by the vehicle live loads. Heavy trucks employed for testing purposes at the AASHTO Test Road in Illinois in the early 1960s were responsible for ejecting large quantities of aggregate, some pieces inches in size, outward to the shoulders of the test road. This action can produce large voids beneath the pavement, eventually leading to collapse of the pavement.

Frost action is of special concern to those involved in route construction, maintenance, and infrastructure repair. With buildings, the solution is rather simple: We just place all exterior footings beneath the frost line, a depth that is known for any given region, and is usually specified in the local building code. For highways, pipelines, and railroads, we do not have so simple a solution. The solutions here are avoidance (of frost-susceptible soils) when possible, or the incorporation of some type of subbase drainage that will prevent capillary rise. The details of such designs are not within the scope of this book.

The most frost-susceptible soils, according to Burmister (1955, p. 144), are coarse silts. This soil size exhibits a substantial capillary rise, combined with a *rate* of capillary rise that is sufficient to foster the rapid growth of the ice lenses during periods of very cold weather.

CLOSURE As may be inferred by the foregoing descriptions of the potential problems with marginally permeable soils, one does not consciously opt to use such soils in modern earth structures if one has such a choice, or if one can afford not to use them. Thus, for example, in Figure 6.1, the only region in which such soils would occur (ideally) would be beneath the dam: the natural ground. The shells and core wall would be compacted free-draining, granular soils (or rock), and properly compacted clay, respectively. The graded filters would be select fills of a specified granular texture. In the real world, of course, we must often make do with whatever materials are available. With respect to dams, we must also recognize that many existing dams were built before the development of rational approaches to design. Thus, there undoubtedly exist many dams whose composition are nothing more than embankments of dumped material of more or less the same material throughout, but necessarily containing enough fines to be rated somewhere between marginally permeable and impervious. In such a case, there is a danger that the phreatic surface (the top flow line of the flow net, Figure 6.1) will extend to and emerge from, the downstream slope, thus creating instability there, including erosion and eventual slope failure. In modern design the porous material of the shell prevents this flow pattern.

6.2.2 Compacted Clays

Compaction curves presented earlier depicted an approximately symmetrical bell-shaped curve (except for select fills), showing a "dry side" and a "wet side." Thus, at any density level, a horizontal line would intersect the compaction curve at two points such as *D* and *W* in Figure 6.5. Since both soils have the same dry density, it might be inferred that both soils have the same engineering properties. For clays, this is not true, and the differences in properties can be significant and pronounced. The differences are caused primarily by the

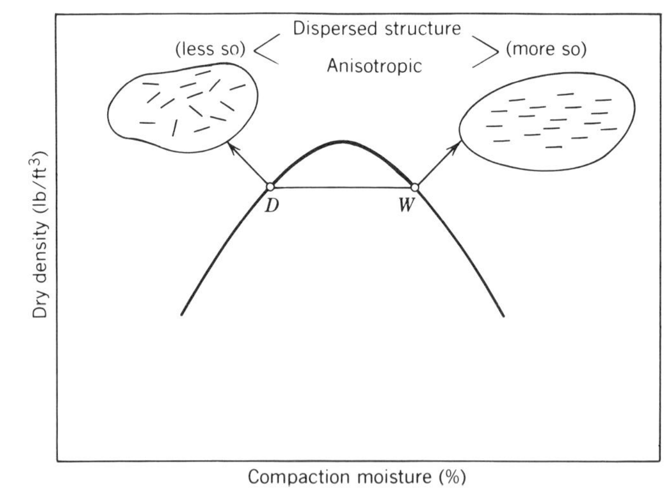

Figure 6.5. *Properties of compacted clays.*

complex *structure* of clays, in turn caused by the size and shapes of clay particles. As determined by electron microscopes, all clay particles are either flat or needle-shaped. Because of their submicroscopic size, the electrochemical forces that develop in their interaction with water are more predominant than gravity (weight) forces. Thus, when a clay is deposited in water, a natural structure develops that includes a complex array of arches among individual particles—a "house of cards" arrangement—termed a honeycomb structure (flocculent in saltwater environments, where the structure is more complex).

When such soils are compacted, there is partial breakdown of the structure toward a more-or-less parallel orientation of the flat (or needle-shaped) particles. This is called a *dispersed* structure. As shown in Figure 6.5 in a schematic way, the tendency toward parallelism is greater on the wet side (*W*) than on the dry side (*D*). In general, it is this difference in particle orientation that causes differences in behavior.

Lambe (1958) has studied the differences in engineering behavior of pairs of compacted soils suggested by Figure 6.5 (*D* and *W*). Following is a brief summary of some of his conclusions:

1. Soils compacted wet of optimum *W* are more anisotropic than those compacted to the same dry density on the dry side (*D*). Anisotropy refers to materials that have different properties in different directions.
2. As a corollary to this, and referring to Figure 6.5, it is evident that soil *W* has higher permeability in the horizontal direction than does soil *D*.
3. Soil *W* has lower shear strength in a horizontal direction than soil *D*.
4. Soil *D* has higher shear strength, in general, than soil *W* in the as-compacted condition, since it has a lower moisture content.
5. However, soil *D* also has a greater potential for expansion upon wetting (a potential negative characteristic).
6. Analogously, soil *W* is weaker, more compressible, and has less potential for expansion than soil *D*.

When writing specifications for the compaction of clays, then, the designer often calls for compaction to be performed 2 or 3% "wet of optimum," or "dry of optimum" depending on the circumstances and function of the fill. For example, designers commonly call for clayey fills that are to support floor slabs or grade beams to be compacted wet of optimum, because clays compacted on the dry side would have the potential to expand upon subsequent wetting and cause damage to the slab or beam. (Of course, a better design would be to call for a select fill for support, but that is not always economically feasible.) As a contrasting example, if one was considering the compaction of the subgrade of a highway, the height of the embankment would be a factor in the decision. For a high embankment, say 20 ft, dry side compaction of the subgrade would provide the greater strength desired, and the potential expansiveness would not be of significant concern because of the high confining pressure of the 20-ft embankment. By similar reasoning, in a region of the highway where the embankment thickness is small, the wet side would be suggested for subgrade compaction.

As a final example, the clay core wall of an earth dam would be compacted wet of optimum to avoid the higher horizontal seepage and concomitant piping potential associated with the more pronounced dispersed structure of clays compacted wet of optimum (see Figure 6.5). In this case, the function of the clay core wall is seepage control; the necessary strength is provided by the shells of the dam.

These examples illustrate the contrasting requirements of strength, stability, and imperviousness.

6.2.3 Select Fills

As described in Chapter 3 and depicted in Figure 3.4, select fills are free-draining soils, predominantly sands. The term *select fills* is becoming more and more meaningful in that soils of such texture are becoming scarcer and thus more expensive. Because of their cost, specifications and designs are often formulated to minimize the quantity needed. Frequently, specifications will call for a pipeline to be seated on a select fill in a backfilled trench, but the fill above the pipe will be the soil that was excavated, assuming such soil is not of disqualifying texture such as peats or highly plastic clays. Similarly, in the case of an earth dam located in a region where select borrow is unusually scarce, a dam may be designed somewhat as shown in Figure 6.6. Instead of having a pervious downstream shell, requiring large quantities of select fill, as in Figure

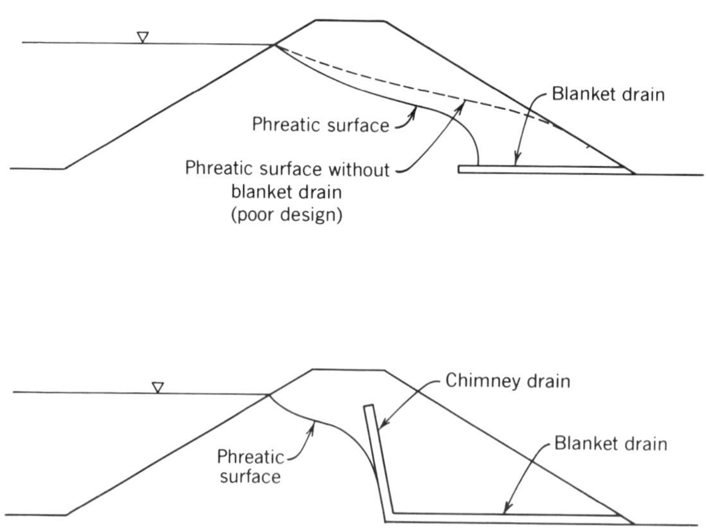

Figure 6.6. *Minimizing select fill in an earth dam.*

6.1, an outlet blanket is provided, and the flow (phreatic surface) is thus contained safely within the dam, as shown. In some cases, the blanket is extended to, and even up, the face of the clay core wall: a chimney drain.

Another use of select fill is the *graded* filter shown in Figure 6.1. It is a basic tenet of good seepage design to provide a gradual transition for the flow of water. In Figure 6.1, for example, without the graded filter on the downstream face of the core wall, the clay would be washed out into the large pore spaces of the shell, a form of piping that could cause serious deterioration of the core. Cedergren (1990) provides criteria for selecting the texture of a filter material for the desired transition.

Select fills may also be used as structural fills to support foundation elements. When compacted to high relative densities, such fills provide the most permanently stable foundation, since they are not susceptible to subsequent changes that may afflict clays or marginally draining soils. One caution: Since select fills are cohesionless, they must be laterally contained, even when dense, to retain their stability. Thus, one must be cautious when excavating adjacent to a foundation element supported by a granular soil, lest the removal of lateral support cause a lateral flow of the soil from beneath the foundation. Such an occurrence can be both costly and embarrassing.

6.2.4 Rock Fills

A properly placed rock fill will provide the necessary strength, stability, and perviousness for an earthen dam, as indicated by the upstream and downstream shells of Figure 6.1. This assumes that the void spaces between rock particles (cobbles, boulders, blocks) are not filled with finer soil particles that would inhibit flow.

The principal difficulties encountered with the placement and compaction (if any) of a rock fill are caused by the size of particles, their shape and surface texture, and the geologic soundness of the individual rock particles.

The angularity of rock particles will depend largely upon the nature of the parent rock body from which it was excavated. As extremes, rock obtained from a stream terrace would be composed of rounded, smooth cobbles; a "ripped" sandstone would be "slabby" and of rough surface texture; and a rock obtained from blasting in a sound rock body might be "blocky" and angular.

As mentioned in Chapter 5 (Section 5.1.4), highly angular soils (and rocks), *if dense*, are extremely strong and stable because of the interlocking of particles; the problem with angular rock fills is getting them that way. In the case of a slabby sandstone, there is a danger of very large triangular void spaces being created by bridging. Large void spaces can also be formed in blocky, angular rock of generally uniform sizes. In the latter case, there would be many point contacts within the rock mass. As the height of the fill increases, the stresses at these points would increase to very high levels because of the small contact areas. The danger lies in the postconstruction failure of large numbers of individual rock particles at contact points or edges, resulting in major settlements.

The general approach to preventing this is to induce such failures during construction, either by using very heavy rollers, where feasible, and by keeping the rock wet throughout the construction period. (There is one incident on record where an earth–rock dam was completed "in the dry." When water was subsequently impounded, the upstream rock fill settled 10 ft, undoubtedly caused by the weakening of the rock, by wetting, at the myriad points and edges of high stress concentrations, resulting in localized cracking, slipping, and shear failures, the aggregate effect of which was sudden settlement. This in turn damaged the core wall, and the entire dam failed.)

The wide variety of factors that can influence the postconstruction behavior of a rock fill makes it a difficult task to decide which methods of control should be employed for a given job. Not the least of these is the importance of the fill in the contexts of both cost and consequences of failure. Clearly, a large earth–rock dam would warrant much greater control than a parking area for a warehouse facility in the back country. The following suggestions are possible considerations:

1. The evaluation of the proposed rock fill, from the standpoint of geologic soundness of the eventual (excavated) rock particles should be made by those with expertise in this area: a geotechnical engineer with experience in the growing discipline of rock mechanics, or an engineering geologist. Rock weathering, geologic defects, mineralogy, and structure of the parent rock will greatly influence fill quality.
2. In large slabby rock fills, it may be sensible to limit the angle of placement repose of the larger slabs to (say) 15°, thus reducing potential postconstruction settlements.
3. For small slabby rock, it may be feasible to use heavily loaded sheepsfoot rollers to fracture the slabs into even smaller pieces. I have seen this work quite well where the parent material was a badly weathered sedimentary rock.
4. As a minimum control, repeated here for emphasis, the rock should be continually wetted as the depth of fill increases, the purpose being to induce failures at point contacts in a gradual, controlled fashion as the height of fill increases.

6.2.5 Silt Fills

Silts (or essentially silty soils) are, in my opinion, the poorest of the four soil types one might use as a load-bearing fill, in the sense that they add little strength to the soil (as with clay binder), but do impart undesirable marginally draining properties to the soil by clogging the voids of what might otherwise be a free-draining soil mixture of, say, sand and gravel. Thus, the soils would be susceptible to instability associated with seepage pressure (boiling, piping), liquefaction, and frost action (Section 6.2.1). In addition, the soil would be difficult to compact because of its moisture sensitivity (see steep compaction curve in Section 3.3).

While silts are poor soils for fills *used as load-bearing materials*, they are the fills of choice for septic systems, in that their marginal permeability provides the needed filtering (cleansing) action (see Chapter 9).

6.2.6 Embankment Stability

When an embankment is constructed for whatever purpose, a design choice must be made with respect to the slopes of each side. Clearly, steeper slopes are cheaper in that they require less material and construction effort, and in the case of route construction, less right-of-way. But the slopes must also be shallow enough and strong enough that they do not fail. In addition, in the case of retaining structures such as earth dams, they must not fail by sliding (at the base of the embankment) or by being overturned. Nor can they be allowed to fail by internal deterioration processes such as have been described in the preceding sections.

Except for extraordinary events such as major earthquakes, freely draining materials such as rock, gravel, and the coarser sands will not fail in the classic, sudden, dramatic sense, as long as they are reasonably well compacted. Distress associated with such materials generally results from intolerable settlements.

Clay soils and marginally permeable soils can and do fail dramatically, including slope failures, bearing-capacity failures, horizontal sliding, and overturning. More often than not, the final failure is a result of long-term internal deterioration (piping) or some other water-related phenomenon caused by the soil's inability to drain freely and quickly. Actually, slope failures, bearing capacity failures, and horizontal sliding are similar in that they all involve *shear* failure, as illustrated by Figure 6.7.

Perhaps the best illustration of how a slope failure is often a water-related phenomenon is to cite what is generally regarded as the most probable occurrence of slope failure with earth dams—the time period following rapid drawdown. There are at least two types of hydraulic structures where the level of the impounded water drops substantially and rapidly on a regular basis, single-purpose flood control dams, and the reservoir embankments of a pumped-storage hydroelectric facility. With respect to the latter, the rapid drawdown generally occurs daily (Rapid here means a matter of hours.)

What happens is illustrated in Figure 6.7*a*. The water level in the reservoir drops to the lower level. As a result, the weight of the soil, previously submerged and thus effectively having a unit weight of about 60 pcf (Archimedes' principle), assumes an effective unit weight of 122.4 pcf, approximately doubled. However, if the soil is clayey or silty, the porewater pressures existing along the potential (shear) failure plane, do not dissipate for some time after the drawdown because of their slow-draining nature. Since porewater pres-

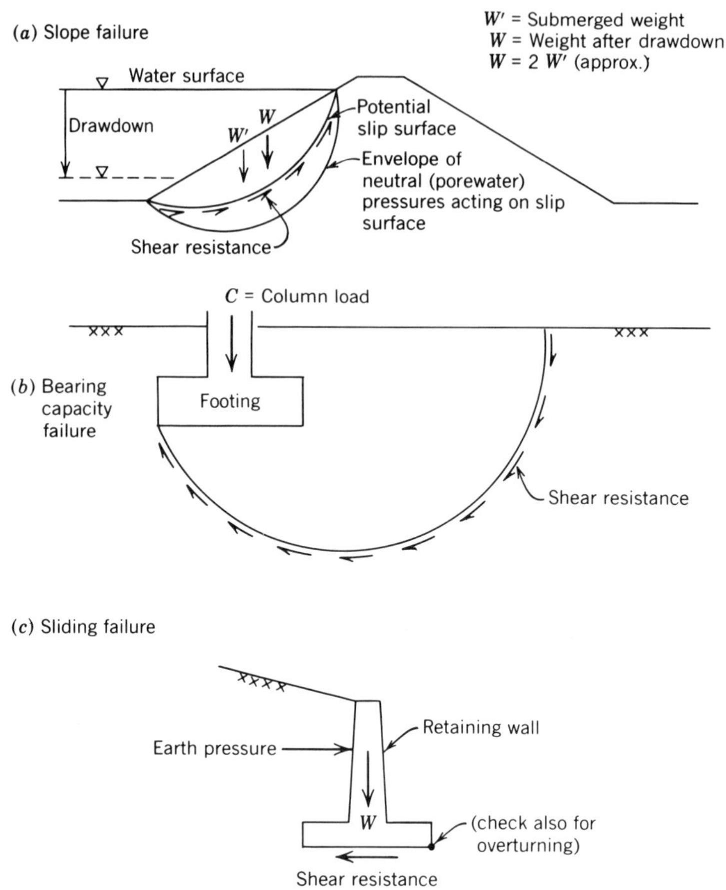

Figure 6.7. *(a) Slope failure; (b) bearing capacity failure; and (c) horizontal sliding: all shear failure.*

sures do not contribute to the effective shear resisting capability of the soil, the potential for slope failure is greatest during this time period, inasmuch as the driving forces (the weight of the soil) have doubled (approximately), but there has been no change in the resisting forces (the shear along the potential failure plane), nor will there be any until such time as the pore pressures dissipate as a result of drainage.

6.2.7 Summary

In this section, I have described briefly an array of problems that can and do occur in earth structures, in the belief that an awareness of such problems will enable designers, contractors, and inspectors to deal with compacted fills more effectively. The interaction of water and soil in marginally permeable soils and the attendant problems of piping, liquefaction, and frost action seem to create most of the problems.

Potential problems with compacted clays, while not as commonly encountered, are nonetheless of equal or greater importance, inasmuch as the failure of earth dams can be and has been catastrophic. Problems with select fills and rock fills are also outlined, and minimum precautions such as vibratory compaction and continuous wetting are recommended. Brief considerations of embankment stability conclude the section.

6.3 CONTROLLED AND UNCONTROLLED FILLS

Throughout my experience in professional practice, and in all my reading about fills, the terms *controlled* and *uncontrolled* fills have been used in specifications, conversations, and the literature, implying that a fill must be one or the other. I recommend, however, that an intermediate category is sensible—a *partially* controlled fill.

It is helpful to one's evaluation of a fill (or any soil or rock, for that matter) to consider two factors—texture and condition. (See Chapters 2 and 3 for fuller commentary on this recommendation.) For fills it is often the case that the texture is controlled to some degree, but not the condition (i.e., placement and compaction control). Simple evidence of this is the frequent signs one sees along highways: Clean Fill Wanted. If such a fill has been in place for many years, its own weight, seepage forces, and frequent trafficking or parking may result in sufficient inadvertent compaction to render the fill quite suitable for light to moderate structural loadings. Alternatively, if the texture is judged acceptable, preferably by test pit inspection, but the condition (density) is not, it may be feasible to compact the soil to acceptable densities by some procedure.

Thus, a fill, once identified as such, should not automatically be rejected for structural loadings just because it is a fill. The important first key is *texture.* Generally speaking, any significant amount of organic material should disqualify the material as a load-bearing fill. Chunks of concrete, or masonry should not be a basis for disqualification—mattresses and garbage, yes!

6.4 NONSTANDARD AND SPECIAL FILLS

The fill types described heretofore have been either of the standard variety (Section 6.2) or of the new, research-oriented type (artificial fills and waste materials, Sections 5.2 and 5.3, respectively). There are a number of fill types that may be classed as nonstandard or special. These include (1) hydraulic fills, (2) sanitary fills, (3) rock–soil fills, and (4) rubble fills.

6.4.1 Hydraulic Fills

Where circumstances and proximity are favorable, a site may be filled by pumping a soil–water mixture to the site from a nearby body of water, typically from a dredging operation of a river or bay (Sowers, 1979, p. 263). Often these bay muds are composed of predominantly silt sizes. Moreover, the economics of the operation are such that loose lifts are several feet or more in thickness. The ideal procedure would be to allow sufficient time for the fill to settle partially under the influence of drainage pressures and its own increasing weight or induce drainage by installing sumps or even well points (at considerable expense). Further compaction can then be accomplished by rolling, by vibroflotation, or by dynamic compaction. (Compaction processes are described more fully in Section 6.5.)

6.4.2 Sanitary Fills

One type of nonstructural fill that is gaining increasing importance is sanitary landfills. I will limit my commentary to listing the requirements stipulated by the United States Environmental Protection Agency (EPA), and will point out the observation by Laguros and Robertson (1980) that there often exist problems of conflicting requirements. For example, the most suitable soil for leachate control (the containment of polluted water) is clay, but it is the worst soil for venting potentially explosive gases.

Sanitary land filling is an engineered method of disposing of solid wastes on land by spreading them in thin layers, compacting them to the smallest practical volume, and covering them with soil each working day in a manner that protects the environment by

1. Preventing the entrance of rodents to the compacted refuse (vector control)
2. Preventing flies and odors from emerging
3. Minimizing the entrance of moisture
4. Allowing the venting of decomposition gases
5. Preventing leachate seepage
6. Inducing the growth of vegetation for final cover
7. Providing a road base to support vehicular traffic

6.4.3 Rock-Soil Fills

On many jobs, any rock excavated at the site is designated as waste in the specifications, thus imposing a double cost in that it is not used as fill and there is a cost associated with its disposal. I suggest that on some jobs it may be feasible and cost-saving to use the rock as fill with alternating lifts of soil. On one job, a rippable sedimentary rock was used in this way, with the added procedure of specifying passes with a heavily loaded sheepsfoot roller over the lifts of slabby rock; the high contact pressures of the roller fractured the slabs of rock and forced the broken pieces into more-or-less horizontal orientation. The subsequent lift of soil (a TALB) was placed with the idea that it would, upon standard compaction, fill the large void spaces of the rock below.

This procedure, while unusual, worked quite well and to the satisfaction of all concerned. It is an example of making do with available materials, a procedure that is becoming more and more necessary as sources of better fills become scarcer. (One colleague recently confided to me that he had to "scratch around" from three or four sources to obtain sufficient borrow for one job. This suggests that the problem of changing borrow (Section 4.3) will become more acute.

6.4.4 Rubble Fills

An assortment of nonstandard materials were described in Chapter 5 and were referred to as either waste materials or artificial fills. These were, and have been investigated as, research projects. One additional waste material, while not generally amenable to research investigations, is demolition rubble. I believe that such materials can serve as an additional source of fill if selected and used in accordance with a variety of common-sense guidelines, many of which have already been presented. Thus, materials such as chunks of reinforced concrete, asphalt, bricks, and stone could serve as acceptable fill components. Techniques of placement and compaction could be the same as described for rock–soil fills in the preceding section. Also, one might consider placing acceptable rubble materials in alternate layers with some other waste material, such as fly ash, to facilitate compaction.

The principal disqualifying factor for the fill material should be based on answers to the following questions: Will the material deteriorate or rot during the projected service life of the fill? Will unacceptably large void spaces be created by its placement? Thus, wood or other organic trash and metal would be excluded.* I would not disqualify reinforcing bars, since much of their surface area would presumably be encased in concrete. Sheet metal is another matter, in that it could create large void spaces by bridging when placed. After rusting, soil from above could settle into the exposed void spaces, probably resulting in significant surface settlement.

6.5 COMPACTORS AND LIFT THICKNESSES

In Section 3.1, I suggested that the most helpful perception of the compaction process was to view it as the expulsion of air from the soil mass (thud-whoosh!). If one considers that a soil can have an *air* permeability, analogous and probably directly related to the more familiar fluid permeability, the relative lift thicknesses of various soil types necessary for efficient compaction become evident. Table 6.1 depicts the approximate range of lift thicknesses for five soil types shown in Figure 3.4; also shown are suggested compactors.

*One exception to this is the increasing successful use of wood chips as fill, as described in Section 5.3.

TABLE 6.1 Lift Thicknesses and Compactors for Various Soil Types

Soil Types	Lift Thickness (in.)	Suggested Compactor[a]
Fat clays	3–5	Sheepsfoot roller
Lean clays	4–7	Sheepsfoot roller
Silts	5–8	Pneumatic (rubber-tired) or tamping-type roller
TALB	8–15	Pneumatic, wobble-wheel roller; smooth drum roller; or smooth vibratory roller
Select fills	24(+)	Smooth drum vibratory roller

[a] For more detailed and comprehensive advice regarding selection of compactors, several organizations have devised charts and tables that list properties and construction characteristics for soils classified according to the Unified System, a system containing 17 categories of soils.

Note. As stated earlier, it is not within the scope of this book to go into details of soil classification. For those unfamiliar with the Unified System, I recommend that you learn the system, or retain a soils specialist to establish the correct classification(s) for the soil(s) of interest. The information may be found in Means and Parcher (1963, p. 119), Ritter and Paquette (1960, p. 206), and in other standard texts under the approximate heading of Classification of Soils. In Chapter 9 a proposal is made for the *exclusive* use of the Burmister system of soil classificaiton, but only time will tell if this recommendation will ever be adopted. If it is, I expect it will be a gradual transition rather than an abrupt change, inasmuch as most practitioners are familiar (and comfortable) with other systems.

A great variety of compactors are available for soil compaction, including different variations of the same class (i.e., sheepsfoot rollers). Fletcher and Smoots (1974, p. 291) and Schroeder (1980, p. 133) describe and present photographs of an assortment of compactors.

Sheepsfoot rollers are available with different foot sizes and shapes, which is probably the most important distinction since the contact pressure is directly related to the size of the feet and, of course, the total weight of the roller, including ballast. (Almost all drum rollers are hollow, so that the total weight can be controlled by the choice of ballast with which to fill the drum—none, water, water plus sand.) When compacting a loose lift of clay, the feet will sink deeply into the soil during the first pass. Upon successive passes, the roller will sink less deeply *if proper compaction is being achieved.* The roller is said to "walk out," compacting from the bottom up.

Compactors that are most suitable for silts or silly soils are either the pneumatic rubber-tired rollers, which provide a kneading action, or a roller of the tamping type, which provides an impact–vibration action. Which type one chooses depends to a large degree on the plasticity of the fines in the soil. Clean silts (those possessing negligible plasticity) would compact best with a vibratory action, but those with significant plasticity (ML in the Unified System) would compact more readily under the action of the kneading of the pneumatic roller. It should be remembered, however, that silts are inherently difficult to compact because of their moisture sensitivity and should be avoided as a choice for load-bearing fills where such a choice is possible.

TALBs can be compacted efficiently with an assortment of different roller types. Again, the choice depends largely on the plastic nature of the finer soil fraction.

Select fills should always be compacted by heavy *vibratory* rollers, with "excess" moisture (flooding), and in lifts of 24 in. or more. I am advised that so-called supercompactors exist that can compact lifts as thick as 10 ft. Only in hydraulic filling operations would one normally encounter lifts of this thickness, and it should be remembered that compaction passes should be made with such rollers only after earlier stage compaction has been achieved by some form of drainage (see Hydraulic Fills, Section 6.4.1).

Occasionally one will encounter a natural, thick deposit of granular soil that has excellent texture but is too loose to support even moderate loading. In such a case, it may be advantageous to employ a technique called vibroflotation. According to Fletcher and Smoots (1974, p. 295), thicknesses of 30–40 ft may be compacted. The vibrofloat is essentially a large vibrating probe that is inserted into the ground and vibrated within the hole while water is simultaneously injected or jetted. As a result the soil "flows" toward the probe, creating a cone of depression much like water flowing to a well during pumping. Sand is more or less continuously backfilled into the depressions surrounding the probe. Soil gradation is an important factor in determining feasibility. It is my belief that angularity of the soil may be of equal or greater importance. (See Section 5.4 for additional commentary.)

For important earthwork projects, vibroflotation might also be considered for the final stage in the compaction of a hydraulic fill.

Another technique for stabilizing thick deposits of soil is *dynamic compaction* or *dynamic consolidation*, a procedure involving the dropping of very heavy weights (12–200 tons) from substantial heights (60–120 ft). This is a relatively new technique, but one that is gaining acceptance. Unlike vibroflotation, it seems to work in soils of clayey texture as well as in granular soils. Loose or soft deposits up to 50 or 60 ft thick have been stabilized successfully. As may be inferred by the numbers cited, the impacts are substantial, and the resultant vibrations may prohibit dynamic consolidation in urban areas. As with vibroflotation, such exotic work requires the careful evaluation of geotechnical and construction specialists. (Because the focus of this book is primarily ordinary techniques of compaction stabilization and the intended audience is mostly nonspecialists, details of vibroflotation and dynamic consolidation are omitted. For those wishing to study these specialized topics fully, I recommend the Geodex Information Retrieval System. I expect that most geotechnical consulting firms and some engineering libraries would have the system and would provide bibliographic information on either—or any—geotechnical subject, at little or no cost. Alternatively, the system itself could be purchased, but it is rather expensive for all but specialists or libraries.)

6.5.1 Making Do

One type of compactor not yet mentioned, principally because it is not normally used for compaction, is the standard Euclid truck, or Euc. On one job where we had to make do in many phases of the job, Eucs were used effectively by loading them with soil and overinflating the tires to about 60–70 psi. (Standard pneumatic rollers go up to about 95 psi.) Although it is a little awkward because of the missed area between the rear tires, the contact width of the four tires was a surprising 8 ft (approximately).

One of my fun discoveries in researching materials dealing with soil compaction was a classic attempt at making do, described in the article "The Uselessness of Elephants in Compacting Fill," by Richard L. Meehan in the *Canadian Geotechnical Journal*, September 1967. One of the more interesting findings of this "modest field investigation" conducted in Thailand, was that an elephant quickly learns to retrace his (her?) steps, avoiding the softer uncompacted areas by exploring the terrain ahead with its trunk (a remarkable sensory organ, according to Meehan). Thus, uniformity of coverage cannot be accomplished readily. I guess one can also conclude from this that elephants are smarter than sheep—which I always suspected.

6.5.2 Confined Areas

I recommend the following experiment to all who read this book. Take a supply of sand, dry it in an oven, and pour it into a 1000-cm^3 graduated cylinder, using a miniature tremie fashioned from a plastic cone and a fitted plastic tube. Fill the cylinder without free-fall (to simulate dumping a fill) to somewhere around the 950-cm^3 level, and record the level. Optionally, place a cylindrical weight on the soil to provide a static confining pressure, and note that the volume change will be negligibly small—even under substantial static pressures. Now strike the cylinder repeatedly with a hammer for about one minute. You will observe that the volume decrease will be about 10%.

This simple, inexpensive experiment illuminates the most problematic of all filling operations: the all-too-common lack (or elimination) of compaction in confined areas, most notably trenches and behind basement walls and retaining walls. For a variety of reasons, it is perhaps understandable that this is often the case. Compaction in confined areas is inherently a labor-intensive operation, and we live in a society where unit labor costs are usually high. In addition, proper compaction is often time-consuming, so the total costs increase accordingly. Finally, because the amount of backfill is usually small, backfilling without proper compaction can usually be done rapidly.

However innocent the practice of dumping backfills may appear, the consequences can be severe. The following case history from my consulting experience illustrates what I mean.

A TRENCH BACKFILL A lateral utility trench on a highway was alleged by an attorney to have created a traffic hazard and caused his client to lose control of his car, resulting in a crash and serious personal injury. I was retained to review the situation, prepare a report of my findings, and provide expert testimony if needed. Investigation revealed that the 7-ft-deep trench had been backfilled by pushing the fill into the trench with a front-end loader, and compaction was achieved "with passes of the wheel of the loader" (a quotation from the contractor's own notebook). The fill "was of a granular texture." With this minimal, but very meaningful, information established, the computation of an approximate maximum potential settlement was made, as follows:

1. Assume that the relative density D_R of the dumped fill is 40%.
2. Assume that the maximum and minimum values of the void ratios for the granular fill are 0.8 and 0.5, respectively.
3. Assert that the documented method of compaction had no significant effect on the densification of the fill (i.e., was totally inadequate).
4. Assume that the settlement is given by

$$\Delta H = \frac{H\,\Delta e}{1 + e_0} \tag{1}$$

where Δe is the change in void ratio, e_0 is the original void ratio, and H is the thickness of the fill. The relative density is

$$D_R = \frac{e_{max} - e_0}{e_{max} - e_{min}} \tag{2}$$

Substituting numerical values and computing e_0 (in this case, the dumped void ratio)

$$0.4 = \frac{0.8 - e_0}{0.8 - 0.5}$$

$$e_0 = 0.68$$

Thus, $\Delta = 0.68 - 0.50 = 0.18$. That is, the dumped soil of $e_0 = 0.68$ would have its void ratio reduced to (possibly) its minimum value $e_{min} = 0.5$, mostly by the influence of traffic vibrations and impact. Computing the settlement,

$$\Delta H = \frac{(84)\,(0.18)}{1.68}$$

$$= 9.0 \text{ in.}$$

Such a settlement could occur (given enough time) in the trench beneath the pavement. With such a deep void space, it is likely that the pavement would collapse into the hole created, and a serious traffic hazard would indeed result.

Observe the correlation between the little experiment and the computations; thus the 10% figure is a nice round number to keep in mind when estimating the potential settlement of a dumped, granular fill. (Note, finally, that the analysis would not pertain to nongranular fills because the range of void ratios for clays and clayey soils is usually much greater. Also, clay fills will settle more readily under the influence of static loads rather than vibration. Shocks, however, can cause flocculent or honeycombed clays to collapse very suddenly.)

RETAINING WALL BACKFILLS A second important type of confined area is behind retaining walls, including basement walls of structures. Ralph Peck and H. O. Ireland (1957) published a paper entitled "Backfill Guide" in which they state,

> It is unfortunate that the problems of working in confined space should occur where the need for good fill is the greatest. The cramped working space, the relatively small volume of fill involved, and the backfill material locally available are undoubtedly responsible for much lax enforcement of specifications.

This repeats and augments some of what I have said. Peck and Ireland also present some instructive and revealing calculations for lateral earth pressures that clearly illustrate the importance of both fill texture and compaction. For dense sand, loose sand, and soft wet clay (or silt), the pressures behind a 15-ft retaining wall were, respectively, 1.69, 2.70, and 5.62 tons/lin ft; for a 30-ft wall, 6.75, 10.0, and 22.5 tons/lin ft. From these data, it is evident that good texture and proper compaction (plus good drainage) have a strong influence on design of subsurface structures and specifications for selection and placement of fills. Enforcement of specifications is no less important.

A few additional observations:

1. If you apply the 10% criterion suggested by my experiment and supporting computations to the 15 ft of loose sand, it does not bode well for an outdoor patio.

2. If a wet clay is used as backfill for a basement wall—as all too often happens—the following scenario may (and did) occur. (This is one of my favorite case histories, as related in a lecture by the late Jacob Feld.) After placement, there was a period of very dry weather, causing the clay to dry and shrink away from the wall. Thereupon a late afternoon shower occurred, followed by a sudden temperature drop. The accumulated rainwater in the space between the clay and the wall froze, exerting pressures on the wall sufficient to push it over. This simple case history, in addition to being a nice construction version of Murphy's law,* graphically illustrates two important facts: the poor quality of clay as a structural fill and the importance of drain-

*If something can go wrong, it will.

age. Not only should a fill be free-draining itself, but water should not be allowed to accumulate behind the wall. In the case of a basement wall, a drainage system must be incorporated in the design that will divert water around or away from the wall. For a retaining wall, suitably spaced weep holes are placed to allow water to flow through the wall.

COMPACTORS AND COMPACTION TECHNIQUES It is evident that the compactors listed in Table 6.1 cannot generally be used for compaction in trenches and behind walls. Instead, one must use hand-operated tampers, most of which are activated by air compressors or internal combustion engines. Fletcher and Smoots (1974, p. 294) and Sowers (1979, p. 254) describe common tampers. As would be expected, the lift thicknesses would be necessarily smaller than those suggested in Table 6.1, generally of the order of 3–6 in. Unfortunately, there appears to be little information available on how much tamping is required for each layer.

Similarly, there is a dearth of information on compaction moisture. As emphasized in Chapter 3, the optimum moisture content for a soil is a function of the energy of compaction. It follows that if we don't know what energy we are using, we cannot stipulate moisture levels with any degree of specificity. One thing that should *not* be done, however, is the practice of jetting or flooding in attempting to compact soils in confined areas. At first reading, the statement may seem contradictory to what I stated in Section 3.3.3 where flooding was generally recommended for select fills. The key distinction here is that we are now referring to compaction techniques in confined areas, more than likely confined in such a way that water cannot readily drain away, because the boundaries of the confined area prevent such drainage. Jetting, flooding, or whatever you call it, will result in the entrapment of water by the fill boundaries (select fills in trenches, for example) or by the fill itself, (clays, silts), resulting in inefficient compaction because of submergence, weakening of the resultant fill, and the potential increase of lateral pressures on walls. I like V. J. Brown's characterization (1967, p. 69):

> [The entrapment of water] would be the same as if we had a huge rubber balloon filled with water beneath the backfill. A vertical load placed on that balloon will transmit equal forces in all directions.

Brown also states, and I concur,

> Too many 20-year and 30-year old specifications that allow flooding still govern trench backfill work, particularly for municipal and for subdivision construction.

The key to whether flooding will aid compaction will be an affirmative answer to the following two questions: (1) Is the soil free-draining? (2) Will the water readily drain through and away from the fill area? Generally, this would apply only to an "open" embankment of select fill. In such a case, the combined effect of flooding and heavy vibratory rolling produces the desired compaction result.

A new machine is now available for compaction in trenches and other confined spaces, called AMERAMAX P33/24. It is a trench compactor that has a four-drum drive vibratory roller, with interchangeable 33 or 24 in. drum widths (smooth or padded). It is completely hydraulic, delivers 13,500 lbs. of force, and features manual and wireless remote control. Address: AMERAMAX, P.O. Box 15759, Durham, NC 27704-0759. Phone 1-800-MIRAMAX.

CLOSURE From what I have described in this section, it is evident that proper compaction in confined areas is a subject that has been badly neglected. To the extent that I have been able to discover, very little research has been done. I expect that this is true because full-scale field testing would be required, requiring expensive excavation machinery and a broad assortment of tampers. The time investment for a well-controlled project would also entail additional major costs.

The one research project of which I am aware is described by Brown (1967) in "Soil Compaction in Narrow Places" where one type of impact compaction was investigated. The Arrow hammer was used for the study. The hammer, with its compaction foot, weighs about 1350 lb and incorporates up to a 9-ft drop. This is clearly not a hand tamper! And the projects for which it would be suited are much larger than the trenches, basement walls, etc. that I have in mind—those tens upon thousands of smaller jobs that would benefit from what we could learn from a carefully controlled research project. I close this section by suggesting that such a project could be funded by a consortium of manufacturers of hand-operated tampers. The work could be done at a large university by a team of researchers under the supervision of a senior faculty member with extensive appropriate construction experience.

Finally, I suggest consideration of the use of artificial fills in confined areas as described in Section 5.2 and illustrated typically by Figures 5.1, 5.3, and 5.8. A specific type of artificial fill of which I have become aware is called Poleset, a pourable, quick-setting, plastic foam. It was considered for a weight-credit backfill application in New Jersey. The unit cost of such fill materials is quite high, so it would not routinely be used in place of soil backfill, but where weight credit is a necessity, the design choice may be viable.

6.6 ENERGY AND MOISTURE CONTROL

As has been emphasized, the approach to controlling field compaction of soils is to test those soils in the laboratory in such a way as to simulate field conditions to the extent that is feasible, technically and economically. Sometimes we violate this for the sake of simplicity. Certainly the impact of a 10-lb hammer on a soil contained in a rigid steel mold hardly simulates the effect of (say) a pneumatic roller passing repeatedly over a soil, but it might be said that we compromise by searching for a bottom-line simulation, irrespective of methods. That is, if the density achieved in the field matches that of the specified target value obtained from the laboratory test, we have simulation.

The combination of factors in the field that will achieve our ends is almost limitless, and it would be difficult if not impossible to state which combination is best. But we can observe some guidelines that lead us to a reasonably workable selection. Table 6.1 lists suggested compactor types and lift thicknesses for various soil types. Once a particular compactor type is chosen, its total weight can be adjusted by choosing the ballast. If we wish to compact to very high densities—say 98% modified Proctor—we would choose the largest compactor type and load up with, for example, sand and water, plus metal ballast. We might also favor lower ranges of lift thicknesses and a larger number of passes. On the other hand, lighter compaction requirements for "unimportant" fills calling for 95% standard Proctor might be achieved by a lighter roller with no ballast, thicker lifts, and fewer passes.

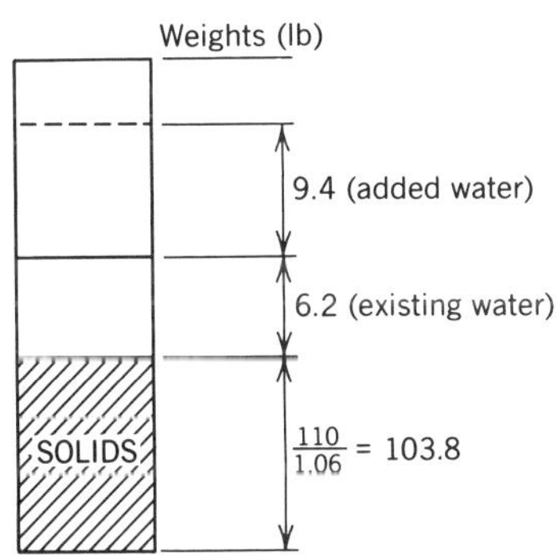

Figure 6.8. *Computing added compaction moisture. A borrow has a unit weight of 110 lb/ft^3 with a moisture content of 6%. Compute the amount of water that should be added to raise its field compaction moisture content to its optimum of 12%. Assume 3% allowance for mixing loss.*

Desired moisture content, w = 15%.

$$\text{Weight of water, } W_w = 0.15 \times 103.8 = 15.6 \text{ lb}$$
$$-6.2$$
$$\text{Added water} = 9.4 \text{ lb}$$

$$\frac{9.4 \text{ lb}}{\text{ft}^3} \times \frac{27 \text{ ft}^3}{\text{yd}} = \frac{254 \text{ lb}}{\text{yd}}$$

or

$$\frac{254 \text{ lb}}{\text{yd}} \times \frac{\text{gal}}{8.34 \text{ lb}} = \frac{30.5 \text{ gal}}{\text{yd}}$$

Compaction moisture levels depend on the energy of compaction; the higher the energy, the less the moisture (see Figure 3.4). The addition of moisture in the field, when necessary, is usually effected either by a truck with a sprinkler bar or by the compactor itself, some of which are equipped with this capability. For compaction to be effective, it is often necessary to allow time for the water to soak into the loose lift of fill, or physically to mix the added water with a grader or bulldozer. Figure 6.8 illustrates a typical computation for determining the desired optimum. If the field humidity is very low, and especially if mixing is thought necessary, it is advisable to overshoot in the computation to allow for evaporation. A 3% overshoot is suggested, but field checking is recommended if deemed important enough.

If the borrow is too wet, the soil may be dried out somewhat by aeration by scarification, again using a grader or bulldozer. I discuss various other alternatives more fully in Chapter 8, Fill Control Procedures.

6.7 GLOSSARY

Capillarity The movement of water (usually thought of as upward movement) in finer-grained soils. Such movement is a major contributing factor to damage caused by frost heave.

Darcy's law The velocity is proportional to the hydraulic gradient; the constant of proportionality is called the coefficient of permeability. Thus, $v = k\ i$. (See Section 6.2.1.)

Flood control dams (single purpose) While many, or perhaps most, dams are multiple-purpose (flood control, water supply, hydroelectric, recreation), if a dam is a single-purpose, flood control dam, its only purpose is to temporarily detain flood waters to protect downstream areas during and shortly after major storms. In such a case, the filling and emptying of the dam is common; hence, rapid drawdown is usual. Rapid drawdown would also commonly occur in levees as flood waters recede. This partially explains why so many earth levees failed (and continue to fail, as I write) during the rise and fall of flood waters of the Great Floods of 1993 in the nation's midwest. Undoubtedly, piping and other forms of earlier deterioration also contributed to the many failures, as described in Sections 6.1 and 6.2.

Grout curtain A wall (or curtain) formed by injecting grout (a thin cement) into porous areas, thus (hopefully) sealing the area and diverting the seepage, as with a sheet pile.

Laminar flow Flow of very low velocity, such that Darcy's law pertains. (Except for most unusual cases of rock fissures or "open gravel," laminar flow would be the case in flow through soil or rock.)

Piezometer For purposes of this chapter, it may be thought of as simply a hollow tube inserted in the soil to specified points for the purpose of monitoring pressure head (e.g., p_Q, Figure 6.2).

Pumped storage A hydroelectric generating installation using two closely spaced water bodies at appropriately different elevations for the purpose of generating electricity. This is done by pumping water from the lower body to the upper body during off-peak (overnight) hours, so that the water will be available at the upper location for peak (day-time) demand requirements. The two bodies are connected by a large-diameter conduit called a *penstock*, with turbines located at the lower water body. The typical daily exchange of water creates daily rapid drawdown. The water bodies may be natural or man-made (reservoirs) or a combination of both. The laws of supply and demand make this relatively new system economically viable.

Sheet pile Pile shapes that are driven and interconnected in such a manner as to create a subsurface wall, in seepage design for the purpose of forcing water to flow under the impervious wall thus created.

CHAPTER 7

Compaction Specifications

SOIL TYPES; STRUCTURES TYPES These appellations originated in the early 1960s through some good-natured banter that I engaged in with an academic colleague—a structures type—who, like me, was oriented more toward professional practice than university research. I used to get on him about the occasional phone calls I'd get from structural engineers that would usually be limited to a statement and a question: "I've got this sandy soil, maybe a little gravel. . . . What's the bearing capacity?" (I was always tempted to suggest that they call Karl Terzaghi.) My colleague, who taught structures to the same students to whom I taught soils, was fond of retorting with: "Wow, it must be neat [we talked like that in those days] to be able to come to class with just a pocket penetrometer!"

As I emphasized in the opening chapters, the focus of this book is on practicing nonspecialists, but includes students and young engineers who may become soils specialists or construction technologists. It is not my intent to presume to advise specialists on how to write specifications; they already know. That being the case, I will concentrate more on how to evaluate specifications others have written, what to look for, what to watch out for. In the process, the potential specialist will also learn to write his or her own specifications.

In Chapter 4, I described some case histories to illustrate major problems in fill control, including those caused by nonspecialists (Section 4.4.1). Thus, the structural engineer who wants to know the bearing capacity is very likely to also dabble in writing compaction specifications.

Morris (1959) has asserted that

> There is a way to write a specification so that the owner will have satisfaction at a reasonable price and the contractor can complete his work with pride and still make a reasonable profit.

I could not agree more.

7.1 TYPICAL SPECIFICATIONS

Studies of earthwork construction practice have shown that most compaction specifications are of either the end-result or methods type, with variations and combinations of both not uncommon. End-result (or performance) specifications stipulate the results to be achieved, usually a target value density as governed by some percentage of a laboratory compaction test. Other end-result criteria include relative density or some type of plate bearing or field penetration test, such as the CBR (California Bearing Ratio). Some percentage of relative density is sensible for select fills, inasmuch as a discrete compaction curve (of the Proctor type) does not exist for such soils (see Figure 3.4). CBR values are normally limited to road bases and subgrades, especially and logically where the controlling agency uses CBR values for designing pavement components. Specifications of the latter type, however, are inherently more expensive, both for the laboratory testing and field verification procedures that are required. The more typical target density requirement may be thought of as an *indicator* of CBR values. This is yet another illustration of the notion of the design flowchart in Figure 1.2.

Methods specifications mean just that. They spell out, sometimes rather precisely, the methods and equipment that the contractor must use, including lift thicknesses, field moisture content, and number of passes.

Sometimes method and end results are specified, a practice that is extremely restrictive on all parties, most notably the contractor. In my opinion, this type of specification is also quite illogical because it is naive to think that a rigid set of methods will always produce a desired end result. When they do not, at the very least it can be embarrassing to the writer of the specifications. Moreover, it will frequently lead to arguments and delays in the field, endless change orders, and often legal entanglements—not to mention a lousy job!

A fourth approach, called the suggested-method and end result (Morris, 1959)

> allows the more experienced contractor the latitude to make use of his experience while it offers a guide to the less knowledgeable contractor. At the same time it insures for the owning agency the desired finished product.

Historically, end-result specifications have been more favored by controlling agencies. In Morris's 1959 survey of the United States, only one state had no requirement for percent density. However, fully 80% also specify "in considerable detail the type of equipment to be used in getting the required density."

In general, it appears that the necessary flexibility is afforded the contractor by the avoidance of imposing other requirements such as lift thicknesses and number of passes. In another more recent international survey (Reichert, 1980), 18 countries reported end-result type, 5 had method types, and 10 had requirements of the mixed type. As stated in Morris's earlier study, the 50 states of the United States each had their own requirements but the majority combined end-result with some form of methods control. In all cases (internationally),

> the contractor is required to take the necessary precautions for preventing the materials employed from becoming water-logged during bad weather. The arsenal of conventional methods includes the provision of slopes for water drainage, immediate compaction of soils after spreading, and surface smoothing of compacted areas.

Readers should keep in mind that the two studies that I have cited here dealt exclusively with roadway embankment compaction. Thus, anyone referring to these sources for guidance in preparing or evaluating specifications should be aware that they do not pertain to structural fills, for example, the support of footings, without perhaps important modifications. In general, but of course depending on the specific loading intensities and other circumstances, the end-result densities one would require would be significantly higher. One should also recognize that there are major differences in the required supporting capacities for roads. Thus, the compaction requirements of the upper few feet of an interstate highway would be more stringent than for the footings of lightly loaded warehouses. Conversely, it is helpful to contrast in a general way the requirements for lightly traveled secondary roads and footing support of heavy industrial buildings or bridge abutments. Thus, the terms structural fill, load-bearing fill, and non-load-bearing have relative meaning only.

One excellent approach to the suggested method and end-result specifications is illustrated by the policy of providing a written suggested guide specification for excavation and placement of compacted earth fill on jobs where such work is a peripheral part of the overall construction. This was the standard procedure of the geotechnical consulting firm where I learned the ropes, Woodward–Clyde–Sherard and Associates. This suggested guide would be a standard document included in all reports to clients, often as an appendix. On jobs where the major focus on the work was to be fill placement, a unique set of specifications would be prepared.

7.1.1 Specification Components

In writing or evaluating specifications the following items should be considered.

1. Stripping and grubbing
2. Excavating
3. Acceptable fill texture
4. Selection of representative samples (for laboratory testing)
5. Fill placement (loose lift thickness)
6. Lift preparation (mixing, moisture adjustments)
7. Compaction equipment (equipment, moisture content, number of passes)
8. End-result stipulation (usually target value density)
9. Weather restrictions and preventive measures pertaining thereto
10. Control (special criteria, procedures, personnel)
11. Correction

STRIPPING AND GRUBBING This work includes the removal of trees, bushes, and topsoil, including roots and other obvious organic matter, such as peats. While such work is an obvious first step for any earthwork project, the extent necessary, as determined by reconnaissance and borings or test pits, can be a significant cost factor for the contractor.

EXCAVATING While excavating is obviously a part of the stripping and grubbing operation, we are here referring to those portions extending to the lines and grades shown on the plans, so the material excavated beyond the organics may or may not be suitable for filling, depending on its texture. Also, the condition and texture of the excavated surface beyond and below the lines and grades may or may not be suitable. Even if borings have been done, surprises often occur when digging commences. Sometimes the excavating itself can disturb the natural condition of the soil. Thus, it is a good idea to include a statement in the specifications designating a qualified person to inspect and approve the excavated surfaces and the excavated materials to make decisions regarding acceptable condition and texture. Unsuitable excavated material would be designated as waste. Alternatively, depending upon circumstances, it may be feasible and sensible to use some marginal materials as fill in some areas (e.g., lightly loaded, parking areas). Excavated soil of stipulated high textural quality would be used as a structural fill or stockpiled for such use.

ACCEPTABLE FILL TEXTURE What constitutes acceptable texture of a fill depends on the intended use of the fill, as has been described rather extensively in Chapter 3 and, especially, Chapter 6. Whether it is a select fill to support a pipe, or a clay for the core wall of an earth dam, the specific textural limitations should be stated in the specifications. The usual minimum specification should refer to grain size limitations for essentially granular soils, typically with respect to limiting the quantities of the very smallest sizes (fines) and the very largest sizes (boulders). For clays, the common basis for defining textural limitations is the plasticity index, but mineralogy and other indices are also used, especially on major jobs such as large earth dams. In such cases, as I have said before and will say again, a geotechnical specialist should be called in.

REPRESENTATIVE SAMPLES A qualified person should be designated to select representative samples from the site or borrow area for purposes of laboratory testing. For textural analysis and compaction testing, this would require about 125 lb of soil taken from the "heart" of the site or borrow mound. If at all possible, a bulldozer should be used to expose a fresh, undisturbed wall of soil from the spot designated by the qualified person. The sample can then be scraped into a bag from a 4- or 5-ft vertical section of the exposed wall. Unless a procedure similar to the one described is used, there is a good chance that the sample will merely be scraped off the ground. Or worse, it will be collected from the base of a slope. In either case, the sample is not likely to be anywhere near representative of the "heart" that will be used in the filling operation, because of the sorting action of rainwater runoff and mass wasting (the accumulation of soil at the base of a slope or cliff, called talus in geologic parlance).

A word about the not-uncommon problem of very heterogeneous soil conditions at the site or borrow area. In such a case, obtaining or even defining a representative sample is an exercise in futility. Until a specific, rational alternate method of establishing field density requirements is developed (such as the Compaction Data Book described in Section 4.3.1), probably the best approach is to specify that an *experienced* soils engineer or technologist be present to inspect and make judgments on field densities. Since the nature of such a job inherently requires many on-the-spot judgments, it is imperative that the person in the field be exceptionally well qualified. The owner should understand, moreover, that certification of the fill cannot be based so much on numerical results as on the judgment of the inspector.

FILL PLACEMENT The loose lift thickness should be specified, and it should be stated that the lifts should be spread horizontally. However, it is often helpful to incorporate some flexibility by allowing changes in lift thicknesses as long as it can be demonstrated in the field that the required target value densities can be achieved.

LIFT PREPARATION Adjustments in moisture content of the loose fill may be required, by aeration (scarification) if too wet, or by spraying with water if too dry. Here, too, rigidity in the specifications should be avoided. For example, the phrase, "the soil shall be compacted at its optimum moisture content" (which I have seen more than once) is both erroneous and—if taken literally—unrealistically rigid. First, a soil does not have *an* optimum moisture content, as implied by the statement, since the optimum moisture content depends on the energy of compaction. Second, the chance that the natural soil moisture will be *exactly* at optimum for the energy of compaction being used is remote.

Thus, if the specification were to be taken literally, the contractor would have to change either the soil moisture or the compaction energy for almost every lift of soil, an obvious impracticality. I recommend that a range of compaction moisture content be specified. In some cases, notably for clays, a limiting lower moisture content might be stipulated to prevent future expansive behavior. Otherwise, let the target value density control. Remember: end result and suggested method.

On some jobs, mixing two soils before compaction may be in order. In this case, and with soil moisture for that matter, avoid the word thoroughly. Be more specific, or at least substitute "to the satisfaction of the soil engineer" (or other qualified person).

COMPACTION While providing suggestions, the stipulation of the exact type of compacting equipment should be avoided. In addition, a range for the number of passes can be suggested, but it is better to stipulate that the methods should be established in the field with regard to equipment, lift thickness, moisture content, and number of passes, specifically in the form of periodic test strips. This is a procedure whereby the inspector and the contractor agree to determine, by a simple trial-and-error procedure, a reasonable combination of methods which will yield the specified end results. Thus, for example, a 150-ft strip of soil can be placed to, say, 9-in. loose lift thickness, and three, five, and seven passes made over three 50-ft lengths. Density tests are made in each section to determine the effects. Thus, early in the job, an agreement is struck on methods to everyone's satisfaction. The use of such a test strip is obligatory in Australia, Finland, and France (Reichert, 1980, p. 188). Incidentally, such a strip can sensibly be within the lines and grades of the plan, to avoid unnecessary extra work. Of course, the methods thus developed will be applicable only if all conditions of the test strip do not change. If the texture of the borrow changes, for example, another test strip and another target value density will be needed.

END RESULT STIPULATION The specifications should call for a very specific end result, usually some percentage of the density established by a recognized laboratory compaction test. A method for doing so is described in Section 4.2.3. Readers should note the limitations of the method, particularly with respect to the soil types to which it applies (generally, TALBs). The extent to which the method has been used in practice is not known, so its performance rating or validity is likewise unknown.* However, the chart upon which the method is based, Figure 4.2, has been used for many years with confidence for the purpose of assigning allowable bearing capacities for footings on sand. A suggested sensible approach for selecting percentage compaction is to increase the value obtained by some arbitrary amount, particularly for fills that

*As noted earlier, Ralph Peck has used a similar method in his practice.

are significantly different in texture from sand. As stated in Chapter 4, the method should not be used at all for clays. For those who use the method with any regularity, field performance records can be accumulated in order to refine and/or gain confidence in its application. The installation of settlement plates, where feasible economically (i.e., if you can get the client to pay for them), would yield the best data on field performance.

Rather than stipulating a single end-result percentage compaction, thereby requiring *all* field tests to meet the stated density, it is often sensible to write the specifications "in terms of an average value [of all tests] and provide two levels, one for its acceptance, guaranteeing the quality of the work, the other for its rejection" (Reichert, 1980, Recommendation 7.3, p. 200). Stating such average values and ranges would seem to be most sensible on very large jobs where wider ranges of borrow texture are likely to be encountered.

Consideration should also be given to stipulating how many field density checks should be made (one test per 3000 yards, for example), which field density technique should be used, and at which locations the tests should be done. Specifics pertaining to these features are treated more fully in Chapter 8.

WEATHER RESTRICTIONS On a fill job of significant size and duration, it is almost inevitable that problems associated with weather will arise. The borrow may be too wet or too dry because of recent rainfall or lack thereof. Cold weather may create freezing problems, particularly for clays or clayey soils. Precipitation during placement may quickly render the soil too wet for compaction. Finally, unless suitable precautions are taken, the compacted fill can be damaged by moisture changes *after* compaction, but often before the full completion of construction.

Wet Soil If the soil is too wet for compaction, as compared to the optimum moisture content for the energy of the compactor being used, the options are to (1) continue operating at less than optimum efficiency, (2) reduce the weight of the roller and use more passes, (3) dry the soil by scarification to an acceptable lower moisture content, or (4) quit for the day or whatever time period will be required for a natural drying of the soils to operable levels.

Which of these options should be followed in a given case and how the decision will be made and by whom are not simple decisions for several reasons. First, there is no exact point at which a soil can be designated "too wet," although obvious undulation of the soil under the action of the moving roller is a clear indication that one is beyond the point of productivity. Second, the decision is clearly an important economic one to the contractor, since all but option (1) will definitely cost money or time, or both. However, option (1) may be the most costly if indeed one is just shoving the wet soil back and forth without increasing its density. Third, who decides? Finally, how does one write the specifications to address the problem?

Since my general recommendation is the use of suggested and end-result specifications, the problem of wet soil can be accommodated by language similar to the following:

> The soils engineer [or other designated inspector, observer] shall advise the contractor regarding suggested options for dealing with the problem of compacting soil which appears to be too wet for efficient compaction.

In this way, the contractor can decide what to do, fully realizing that the end result (field density, usually) will ultimately govern acceptability.

Dry Soil Here again, the designated person can at some point advise the contractor to consider the option of wetting and mixing the soil prior to compaction. Where clay fills are concerned, however, the end result might include the stipulation of a lower limit on moisture content *and* compacted density, to ensure a completed earth structure that is neither too compressible nor detrimentally expansive.

Cold Weather Specifications should prohibit the placement and compaction of soils that contain ice and frozen clumps of soil. This would normally be a disqualifying problem when dealing with soils containing silts and clays of significant quantity. Clean, granular soils such as select fills and sand–gravel mixtures may be placed and compacted in such weather, however.

Precipitation During Placement Steady rain at the site, if not too intense, can often be handled by specifying (or "suggesting") that "in the event of rain, the soil shall be compacted immediately upon spreading."

Postcompaction Precautions The specifications should require the contractor to provide slopes necessary for surface drainage. It may also be advisable to include requirements that will protect against surface damage, notably by freezing, drying, or construction traffic. (In one interesting case, construction workers had warmed themselves with a fire in a 55-gal drum. As still another example of Murphy's law, the spot where they placed the barrel was precisely where a footing was later placed; the baked clayey fill, when subsequently wetted, expanded and cracked the footing, column, and wall above.)

CONTROL In addition to the usual end-result stipulation of a target value density, other features of fill control may sometimes be included in the specifications. These could include special criteria, matters of fill control procedures, and how various persons are (or should be) involved with implementation and enforcement.

CORRECTION The specifications should contain language pertaining to the field testing of the fill and correction measures required where tests indicate inadequate compliance. An example of the latter might be that "any lift that is not compacted to the specified density should be recompacted until the required density is attained."

It may be advisable to include a statement that will have the effect of prohibiting the unsupervised placement of large thicknesses of fill. Obviously, such a feature will make sense only where the contract calls for the continuous presence of an inspector, observer, or other designated person. A clause requiring the contractor to excavate to a level specified by the on-site person should also be considered. This is to guard against "lunch hour filling," a procedure where large fill thicknesses are placed, usually in small, confined areas, and the surface compacted. In such cases, the upper foot or so may meet the target value density, but the lower regions most likely would not. (More about these kinds of sticky problems in Chapter 8.)

Special Criteria When large or important earth structures are to be built, criteria other than specified densities are sometimes used. Usually, such criteria create more expense and are more technically complex, requiring the attention and involvement of geotechnical specialists. Following is a summary of recent practice, most of which is described more fully in the 1980 International Conference on Compaction (Paris, France, three volumes).

Reichert's review of reports from 32 countries (of 67 polled) reveals (Reichert, 1980, p. 181) that

> The four most frequently used criteria are: the imposition of a Proctor density, the limitation of air percentage, the imposition of a CBR value or of a value for the bearing capacity as given by the plate bearing test.

Of these, the Proctor density is most often used (thus confirming the focus of this book).

Japan and New Zealand impose upper limits of air percentage in the compacted soil, and these limits are typically and logically applicable to essentially cohesive soils. France also uses this criterion for lime-stabilized soils. In Chapter 3, I suggested that the best definition of compaction is the expulsion of air from the soil, so specifying a limiting air percentage is really an indirect way of stipulating a degree of compaction. If too much air remains in the soil, postconstruction loadings will result in rather rapid settlements until 100% saturation (zero air voids) is achieved.

CBR values are used as criteria by Swiss and Thai authorities, generally for road bases and subbases. The CBR (California Bearing Ratio) is rather widely used in the design of pavement support systems and so it would seem quite logical to use the test for compaction control when constructing roadway systems. The test involves the laboratory compaction of a representative sample of the soil in a cylindrical mold, a 4-day soaking period under a surcharge weight simulating field pressure conditions, and penetration with a penetrometer of approximately 2-in. diameter. The CBR is the ratio of the measured resistance (usually at 0.1-in. penetration) to that of a standard crushed rock. The test may he conducted in a similar manner on a prepared section in the field.

Reichert makes the pertinent observation, as part of his conclusions (p. 198), that

> the CBR test should be excluded from certain requirements because of the low diameter of the plunger, the result of the test being very susceptible to the least lack of uniformity in the surface (e.g., presence of a small stone).

(One of my favorite stories along these lines is the consolidation test that wouldn't consolidate. After several puzzling days, the ¾-in.-thick sample in the ring was extracted, and it was discovered that a ¾-in.-thick *whole clam* had been neatly encased! Truly a marine clay, if there ever was one.)

Plate bearing tests are used by nine countries. Most commonly, the test is performed for the purpose of determining the subgrade modulus, a measure of the supporting capacity of the subgrade, which is needed to select the thickness of rigid pavements. The test involves the field loading of a 30-in.-diameter plate, with the usual specification that the test is to be made with the soil at or close to its optimum moisture content (Hough, 1969, p. 480). Thus, it is seen that the plate bearing test is not a substitute for compaction testing, but rather an additional control criterion, with, of course, the added cost of the field loading test.

In contrast to the apparent high cost of conventional plate load testing of cohesive soils, a very interesting and promising method of using plate loading tests for controlling compaction of granular soils, rapidly and at low cost, is described by Giddings (1980, p. 547). The method was developed for a large project in South Africa, involving 5 million cubic meters of variable granular material. A conventional tractor was rigged with a front-mounted hydraulic jack and further modified (ballasted) to enable the application of a 2.2-ton load to a test area through a circular steel plate of 30-cm diameter. A guage was used to record settlement to enable the computation of the elastic modulus. Each test required only about five minutes, and the results were immediately available to accept or reject the compaction. Perhaps most interesting is the fact that no laboratory compaction is required. As Giddings asserts:

> In the compaction of earthworks, the essential properties of a soil are those of strength and deformation. It is far more logical to obtain a strength or deformation parameter which has real meaning to the engineer, rather than to measure percentage compaction. Percentage compaction is, in fact, only an approximate indicator of the value of these essential properties, established during many decades of experience in varying conditions.

This is the only instance of which I am aware where measuring engineering properties is apparently cheaper than obtaining index properties, in contradiction to Figure 1.2, an interesting anomaly. Giddings is investigating extending and modifying the method, presently limited to granular materials, to include application to more clayey soils. This would seem to be a very worthwhile endeavor, especially for possible applications to very large earthwork jobs where field research and the development of specialized equipment can be justified economically.

However, a question arises regarding whether or not Gidding's plate loading control method is generally applicable, even to granular soils, specifically because of the static nature of the test. It is well known that rather loose granular soils can sustain substantial static loadings but will settle dramatically under the influence of vibrations. In Section 6.5.2, I described and suggested a simple experiment using a 1000-cm^3 graduated cylinder. After placing sand loosely in the cylinder, I suggested that a static load—even a substantial one—would not induce a significant settlement. Having performed this experiment many times as a classroom demonstration, I can report that the load was my own weight; I rather awkwardly chinned myself on a piston. I then had one of the students confirm that little settlement took place under the influence of the static loading. A hammer was then used to demonstrate the dramatic effects of vibration on the settlement of the sand (about 10%). I have computed the static pressure in the cylinder experiment and, by a remarkable coincidence, the pressure is almost exactly the same as that used in Gidding's plate loading test, 40 psi (200 lb/5 in.^{2}; 2.2 tons/109 in.^{2}).

Figure 7.1 illustrates what appears to be a contradiction to the assertion inferred by my cylinder experiment—that the settlement (or elastic modulus) should not be significantly affected by the density of the soil in a static test. I believe the apparent contradiction can be explained on the basis of differ-

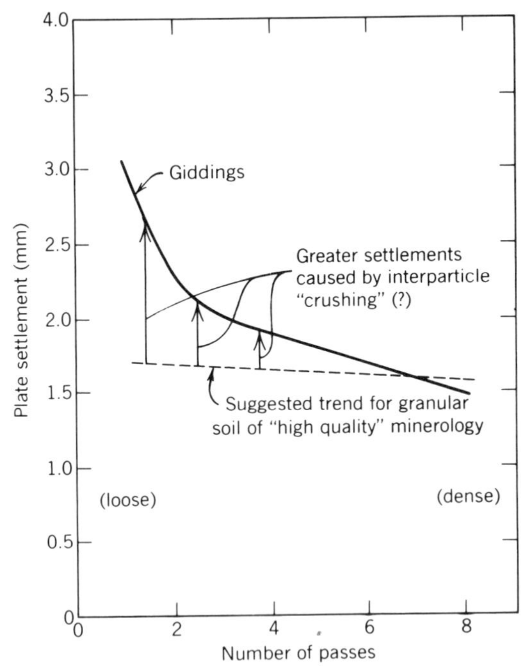

Figure 7.1. *Giddings' plate loading tests. (After Giddings, 1980.)*

ences in soil mineralogy. The soil in the South African project is described by Giddings as of "a porous nature . . . of weakly cemented material . . . [which would be] broken down by laboratory compaction." The field compaction was by "very heavy vibratory compaction machinery." For the plate loading tests of Figure 7.1, compaction was done by "a towed single drum vibrating roller of 14 tons static mass" (Giddings, 1980).

What I believe happened was that the static loading of the plate hearing test induced structural breakdowns of the weakly cemented soil particles at the intergranular contact points within the zone of influence (pressure bulb) of the loaded plate, most of which occurred when the soil was in the relatively looser conditions (lower number of passes). In the case of the cylinder experiment, admittedly a very crude test, the loose high-quality sand was able to resist the static pressure without interparticle breakdown, and no significant settlement occurred. (The sand used in the cylinder was Ottawa Sand, the same type used in the conventional sand-cone field density test.)

If my speculation is correct, a similar field experiment on a high-quality granular material (of sand sizes) would result in a much shallower curve, irrespective of density, and that even a loose soil would give the illusion of denseness because of the static nature of the plate loading test. Fortunately, the construction method involved six passes of a heavy vibratory roller, so the quality of the completed earth structures was assured. My question concerns whether or not the method of control applies in general to the compaction of granular soils, for there would be a danger of accepting loose soils based on static testing, only to have postconstruction vibrations induce intolerable settlements. There is a classic, famous case history that illustrates this concern. Terzaghi was called in to investigate the settlement of a series of structures that had stood for many years without movement or distress. With his unparalleled skills of observation and of assessing cause and effect, he determined that loose sands underlying the row of structures had started to settle under the new influence of heavy truck traffic on the road upon which the structures stood. The clue was that all of the structures were tilted toward the road. Apparently the vibrations were sufficiently damped so that there was little effect on the sand toward the rear of the structures.

Hendron and Holish (1980, p. 565) report a method of compaction control of a major earth dike system of 3 million cubic yards using cohesive soils with widely variable properties. Their approach was to augment density testing (laboratory and field) with the development of correlations between direct measurements of degree of compaction and various index properties of the soil, including liquid limit, the fines content, and the unconfined compression strength as measured with a pocket penetrometer. This major field effort was interestingly similar to the approach I describe and suggest for the problem of "Changing Borrow" (Section 4.3).

Eggestad (1980, p. 531) describes a method of compaction control that he developed and field tested called a comprimeter, a penetration apparatus based on the common-sense principle that a rod driven into a dense soil will cause heave of the surrounding surface, and that the heave volume would be directly related to the degree of denseness (including volume reduction for very loose soils). The device was compared to the more common control methods of field-density testing: the water-balloon and nuclear device methods. The results showed that the most rapid and simple method—the comprimeter method—gave very reasonable results for sands and sand-gravel mixtures. Each field measurement took less than 5 minutes. In his conclusions, Eggestad makes the salient observation that "compaction, even in small earthworks should be controlled."

Finally, Reichert reports that other control methods are sometimes used when conventional tests are not feasible; for example, proof rolling of rock fills where the sizes of the rock preclude tests such as CBR or plate bearing.

Fill Control Procedures Depending on circumstances, and perhaps especially on how far along the project has been developed in terms of planning, it may be advisable to include some language in the specifications covering fill control procedures. As stated under the end-result stipulation, the number of field density tests and methods for controlling (choosing) their locations should be stated. But also included might be such things as methods of testing: field-density testing (water balloon, sand cone, or nuclear), tests for texture (sieving, plasticity), or the specifics regarding special field testing criteria such as CBR, plate loading, or full-scale tests for special circumstances.

Because one of the purposes of this book is to provide what amounts to a manual for such work, the full details of these procedures are consolidated in the following chapter.

7.2 IMPLEMENTATION AND ENFORCEMENT

We come now to a very difficult and sensitive aspect of fill control or, indeed, any type of engineered construction: How do we go about implementing and enforcing the features of the contract documents, including the plans and specifications? Corollary questions would include: Who has the responsibility? The authority? Who gets the blame when something goes wrong? Easy questions, tough answers!

Aggravating the problems considerably are some factors that have evolved and become disturbingly pronounced over the past 20 or 25 years. One is a broad, national trend in American life. The others are direct results, I believe, of the national trend.

We have become a litigious society. The number of American lawyers increased by 83% in the 1970s. By contrast, Japan produces twice as many engineers as we do with half our population base. In fact, there has been a decline of between 5 and 10% in American scientists and engineers during the same time period. Now, the American Bar Association and the American Civil Liberties Union may not perceive these data in the same light as those involved in engineered construction, but no less an authority than the former Chief Justice of the United States Supreme Court, Warren Burger, has spoken out against the proliferation of lawsuits and the resultant backlog in the courts. Certainly there are many good and proper instances where a person needs legal recourse to resolve a problem, but it would appear that a serious imbalance has developed.

One incident from my own experience comes to mind. I had been retained by a law firm as a potential expert witness dealing with a foundation failure. Before the case got to trial, during the deposition phase, I met with a lawyer from the firm that retained me. He confided to me a modus operandi whereby one cites as many defendants as one can possibly identify, however remotely involved with the construction failure, so that there will be more potential contributors to the settlement pot. (Negotiations for a monetary settlement go on continuously before and even during the trial.) Upon later reflection on this revelation, I realized that I could be one of the defendants cited in some future lawsuit, particularly if the modus operandi is indeed common practice. The insidious closure to this loop is that you would have to hire a defense attorney, and the expense may be more than it would cost to contribute to the settlement. Thus, a completely innocent person can be damaged financially and his reputation tarnished by the stigma of settlement.

A further effect of this litigiousness is reflected in the decision of an academic colleague of mine to decline part-time design consulting work because of a legitimate fear of being dragged into a lawsuit in the manner described. The cost of professional liability insurance has gone up in accordance with the litigation explosion, and the small volume of work that a typical engineering professor has time for does not generate sufficient income to warrant the expense of the insurance. The man in question is a highly competent, well-educated, licensed, professional engineer (M.S., structures; Ph.D., geotechnical; postdoctoral, Fulbright). Thus, the public is partially deprived of the contributions of one of its better engineers. For many years I did small consulting jobs without liability insurance, figuring my competence would serve me well enough as protection. I now see that that is not necessarily the case. I know of other part-time consultants who operate without insurance, but as they say in Atlantic City: "It's a crap shoot." Of course, full-time practitioners operate with full insurance protection for themselves and their employees. If they don't, they'd better! These costs are, of course, passed on to their clients. Joe Public always pays the bills!

If any further proof of this national trend is needed, consider this assertion: "As a field engineer, you *will* appear before a court of justice during your career!" This quotation (emphasis mine) is from the first chapter, first page, of a book entitled, *Sue the Bastards: Handbook for the Field Engineer*, by Frederick Richards, P.E., 1976. The paperback was sponsored by the Rochester Section of the American Society of Civil Engineers. The startling cover, which goes along with the graphic title, is a photograph of a hard-hat worker in a jail cell! Interesting if a little scary reading.

The effect of litigation fear (as Woody Allen says, "you're not paranoid if people are *really* out to get you.") in field work is the introduction of a new breed of field worker: the *observer.* I do not know when this person came into being, for there were none in the early 1960s when I was engaged in fill control work as an *inspector.* I suspect that some bright lawyer made a convincing semantic link between inspection, responsibility, and culpability, won the case, and thus established a legal precedent along those lines. The response was the birth of Omar, the Observer. I have chosen the name of this fictitious person for two reasons. First, it is alliterative, and I've always liked that. Second, I don't know anyone by that name, and I'd hate to lose a friend or even offend a casual acquaintance. For Omar is somewhat scatterbrained. He is the fellow about whom everyone asks, "Who hired him, and why was he hired?" (More often than not, he was somebody's nephew.)

Mostly, I heard about Omar through comments by graduate students who were working full time as engineers in training. Lacking first-hand experience with this development, I have talked with colleagues in professional practice who have dealt with this new feature of construction practice at the management level and have solicited their opinions of its ramifications. That is the basis of what I report to you here.

What I suspected, and has been confirmed, is that observers serve only to exacerbate the problem, adding more confusion and more cost to an already confused and costly situation, principally by making it more difficult than ever to establish clear areas of responsibility and authority. As T.B.* put it, "Anyone who accepts responsibility without authority is crazy."

To the extent that the term inspector may suggest authority, the designation observer seems to connote impotence. In a hypothetical but potentially very real scenario, an observer might report. "I told him the fill was no good, but he poured the footing anyway. "T.B. would respond, "What do I need you out there for; I could put in a television camera."

What it amounts to is that, somewhere along the line, the buck has to stop, sensibly with someone in the controlling agency. For example, a professional engineer will have to state in writing that "the specifications for the fill have been complied with" and affix a seal accordingly. It would seem that the use of the term observer in engineered construction of any sort is a disservice to all involved. First, as I have noted already, it is common that young, relatively inexperienced people are currently assigned to fill control field work (although this may change with the advent of construction engineering technology curricula). To further saddle them with the designation observer is, in my view, ill-advised, for it will further reduce their potential for effectiveness from a tenuous position to one of virtual powerlessness. The client, of course, will not be happy with prospects for getting value for his money. The engineer, or whoever must certify the work in writing, will not be able to do so confidently if the field person has not been able to exercise any authority on a day-to-day

*See acknowledgments.

basis. And finally, the contractor will eventually suffer the inevitable consequences of performing without the daily pressures of reasonable and effective control procedures. (While some contractors may scoff at this, I am convinced that the best chances for success and profit spring from long-range reputation for the quality of completed structures that are free of postconstruction problems. I further believe that it is the nature of us all that we need a *reasonable* degree of steady pressure to operate effectively—that includes inspectors, engineers, and contractors.)

I would thus discourage the use of the term observers in contracts and specifications. Instead, I would encourage the return to the use of terms such as inspector or controlling agent, and strengthen the line of control and authority by inserting appropriate language in the contract documents, language that would carefully define responsibility and authority. For fill control work, I suggest the following as a guideline for terminology and tone:

> The inspector shall advise the contractor of the rejection of any unacceptable end results, reiterate steps toward correction as may be stipulated in the specifications, and (optionally) advise the contractor on alternative methods of obtaining the end results required.

7.3 NONTECHNICAL ASPECTS OF SPECIFICATIONS

Fletcher and Smoots (1974, p. 390) stated that specifications should set forth all the requirements of the work to be performed. I believe that I have included all the technical aspects of compaction specifications in the preceding sections of this chapter, and that is the stated focus of this book. There are, however, other aspects of specifications, not yet mentioned, which should be included. These include (Fletcher and Smoots) "terms of payment, time allowed for completion, and provisions for adjustment for changes in the work." Such features are clearly of vital importance to the client and the contractor.

7.4 SPECIFICATION AND PROJECT EVALUATION

Since this book is written largely for nonspecialists, but including field engineers and engineering technologists (collectively, the inspectors), most readers would be in the position usually of being given a set of specifications for whatever their various purposes might be: the client perhaps to see what he is paying for, the contractor to bid the job, the architect or structural engineer because of his or her indirect but important involvement with earthwork operations, the construction engineer because of the need to advise the contractor directly, and the inspector because of very direct involvement with control. Thus, the typical need of such persons would be to evaluate specifications, rather than write them, and to otherwise review all aspects of the project prior to the start of construction.

One of the more obvious ways to do this is simply to read the preceding sections of this chapter and make an item-by-item comparison of your specs with what is described here, in a checklist fashion. I recommend this procedure, but would also offer some additional guidelines, for sometimes one can get so bogged down in details that one fails to see the forest for the trees. Group your evaluation process into two categories: soil and equipment.

7.4.1 Soil: Quantity, Texture, and Density Requirements

1. Is the quantity of soil (or rock) sufficient for the volume of fill needed for the site preparation work?
2. Is the texture of the soil (or rock) that is available in accordance with the specifications?
3. What will be the acquisition problems and costs?

These questions must be considered together. If the site preparation involves cuts, as it usually will, will the excavated soil be of sufficient amount and of suitable texture? You'll need to have borings, test pits, or (at the very least) some soil maps describing (probable) soil texture to get the answers needed. Soil maps for most counties of the United States are available through regional offices of the Soil Conservation Service (SCS), a division of the U. S. Department of Agriculture. Maps and accompanying text describe soils' texture to a depth of about 10 ft and also provide information on depth to bedrock, drainage characteristics, and depth to water table. For large jobs, air photos may be very helpful in locating potential borrow sources, especially where soil maps are not available (e.g., outside the United States). Air photo interpretation is the art of identifying geologic landforms through the study of clues such as drainage patterns, gully erosion, tone, vegetation, and land use. Once the landform has been identified, the texture of the soil in the deposit can be predicted within reasonably well-defined limits. A limited number of borings or test pits may be planned for purposes of confirmation. However, other acquisition problems should be considered simultaneously or in some logical, sequential fashion. Who owns the land? Will they sell it? How much? What is the haul distance? Will there be access (trafficking) problems? How much stripping and grubbing will be required?

4. Are the density requirements sensible and specific?

If the job is large and complex, several different types of soil texture may be needed, each compacted to densities compatible with their intended use. The end-result densities should be in terms of specific laboratory test names (standard Proctor, modified Proctor, relative density, and/or their correct ASTM designations).

In going through this evaluation process, the nonspecialist should get a feel for whether the specifications are well written and complete. He or she should get a sense of whether the specifications are what Fletcher and Smoots call "a cut and paste" document from other jobs. Finally, one will more than likely be able to decide whether a specialist's services are needed. If so, this is my best advice: Call one in before it's too late.

7.4.2 Equipment: Excavation, Loading, Hauling, Placement, Precompaction Preparation, Compaction

Is all the equipment available to do the job, and is there a reasonably good balance of the different kinds of equipment? The logic of the title of this section, and the broad question stated above, is based on a consideration of the chronological sequence of operations in any earthwork construction: The soil (rock) has to be excavated, loaded, hauled, placed, prepared (sometimes), and, finally, compacted. As a simple example of the balance referred to in the question, it can sometimes develop that one has too much placement equipment relative to one's compaction capability. (In a case history that I describe in Chapter 8, this happened to an almost laughable degree.)

Excavation can, of course, be accomplished with an extremely wide assortment of equipment, from hand shovels to blasting. Not only is the amount of excavation of concern, but the ease with which it can be done is of major importance in terms of equipment selection and cost. Indeed, I would venture an educated guess that the accuracy of estimating the quantity and ease of rock excavation can, and often does, make the difference between profit and loss on a project. As one contractor stated: "When the estimate of rock is out 50%, that's not too bad. When it is out 100% we can live with it. But when it is out 10,000%—as on our last job—somebody gets hurt" (Greer and Moorhouse, 1968). In addition to determining the amount of rock excavation, the question of whether the rock will require blasting or ripping is a major cost factor, and, of course, the selection of equipment cannot be made without first making this determination. A common means of doing this is the use of a rippability chart based on seismic refraction exploration (Schroeder, 1980, pp. 109–110).

To a lesser degree, the ease of excavation of soils can be important. For example, if all excavation is to be done on or very close to the site, and the soil is loose or soft, one special-purpose machine—the self-propelled scraper—may be used for three operations: excavation, hauling, and placement. (A scraper is a machine designed for the sole purpose of picking up soil from one place and spreading it in a lift where required. It is not an over-the-road vehicle.) If the soil must be hauled from a borrow area remote from the site, involving transportation over public roads, then a wider assortment of equipment will be needed. Even when only on-site excavation is required, however, the scraper may not have the power necessary to excavate soil that is dense and of significant cohesiveness, Glacial tills, for example, typical in northern sections of the United States, are almost always difficult to excavate, requiring power shovels or similar expensive equipment. If self-propelled scrapers are used, a bulldozer will commonly be required to assist the scraper by pushing.

Alternatively, a tractor-drawn scraper may be used. Overlooking this need, particularly on a large earthwork job, could result in significant loss (or reduced profit) to the contractor.

The best type of equipment for placing soils is the scraper. An experienced operator can eject the soil from the "bowl" of the scraper in a controlled fashion to spread the soil in a thin layer over a strip, typically in accordance with the lift thickness stipulated in the specifications or as modified by performing test strips in the field.

Soils may also be placed by some type of over-the-road haul unit (i.e., a dump truck). This would normally be the case when the soil must be trucked in from an off-site borrow area. In such a case, the dumped soil must be spread, usually by a bulldozer, to the required lift thickness before compaction (or any required precompaction preparation).

Tank trucks with sprinkler bars, or some type of equipment for drying by scarification, may be required to adjust the moisture content of the soil.

The final step is, of course, compaction. Guidelines for more detailed evaluation and selection of equipment for excavation, hauling, and placement of soils are provided in a recent, thorough treatment of the subject, *General Excavation Methods* (1980), by A. Brinton Carson, General Contractor and Professional Engineer. To augment recommendations and other sources given in this book, I also recommend Carson's lists and discussions of different types of compaction rollers (Chapter 11).

7.4.3 Compliance with Applicable Laws and Regulations

In evaluating (or writing) specifications, one must be certain that there are no conflicts or omissions with respect to applicable building codes, ordinances, or an assortment of other regulations, increasingly of an environmental or safety nature (Occupational Safety and Health Administration, OSHA).

7.5 GLOSSARY

Elastic modulus The slope of a stress–strain curve. If the material is elastic (as steel), this slope is a straight line. If the material is other than elastic, as with soft soils, an appropriate representation of the stress–strain curve is determined, usually then called a *subgrade* modulus (as in field plate bearing tests).

Geologic landforms A distinctive form, often in both shape and texture, associated with a particular geologic process. Once the landform is identified (e.g., by aerial—or even satellite—photography), much useful information about texture and potential foundation suitability can be predicted. In geology, the study of landforms is called *geomorphology*.

Proof rolling Rolling a soil or rock surface with a very heavily loaded compactor for the purpose of locating loose or soft areas; "trouble-shooting."

Settlement plates Field instrumentation of a completed structure that permits periodic readings, using an engineer's level, for the determination of settlement and settlement rates. Although the term plates is used, the instrumentation can take any form, as simple as a scribed horizontal line on a wall. For fills, a horizontal plate, with an attached riser pipe extending out of the fill, is commonly used.

CHAPTER 8

Fill Control Procedures—Inspections

After all is said and done, the bottom line (as our lawyer friends say) is Will the job get done as planned so that no postconstruction distress will occur? Assuming that the specifications have been well written, this will depend to a large degree on whether the specifications are followed by the conractor. On typical large jobs, where extensive filling is required, a necessary ingredient in compliance is enforcement by the controlling agency, followed by written certification to the owner. Ideally, the controlling agency would be one with expertise in geotechnical matters, such as a consulting firm or a highway department. The direct controlling agent is, of course, the field inspector. This chapter is largely directed to the latter. To a lesser extent, it is also directed to contractors in recognition of the fact that there will be many smaller jobs, or special circumstances, where no inspector will be present, so that compliance with specifications (or good construction practice when there are no specifications) will be based on pride in workmanship and reputation for excellence—a priceless commodity for a contractor (or anyone else, for that matter!).

There is even something here for structures specialists.

8.1 FIELD-DENSITY TESTING

Assuming homogeneous soil conditions, if all footings of a structure have the same bearing pressure, will there be any differential settlements between footings? Think a bit *before* checking the answer! The answer is at the end of the chapter.

8.1.1 The Pressure Bulb

I started this chapter with a quiz to illustrate a common serious misconception, particularly among structures types. Although I have not compiled records, I can say with certainty that a large majority of engineers and engineering students will answer the question incorrectly. I think I know why, and it is partly the fault of those of us who write textbooks. The source of the problem is that bearing capacities are specified in tons per *square* foot (or similar units). The use of planar units leads a reader to think only in terms of the pressure that exists between the footing and the soil—the contact pressure. What governs the settlement, however, is, of course, the *volume* of soil that is stressed significantly, hence the fundamentally important concept of the *pressure bulb.* There are, in fact, an infinite number of pressure bulbs, as revealed by theoretical studies of stress distributions in semiinfinite masses. The ever-pragmatic engineer, however, asks the mathematician, "To what regions do significant stresses extend?" To which the mathematician responds, "What's significant?" Engineer: "Oh, say 10% of the contact stress.

The answer is shown in Figure 8.1, which depicts a loaded area of width B exerting a contact stress p on the homogeneous soil. The $0.1p$ pressure bulb may be thought of as the standard pressure bulb, or better yet, the rule-of-thumb pressure bulb. There are a number of reasons for the latter nomenclature. First, the shape of the loaded area in application would vary, typically square, rectangular, or long for footings, and elliptical for wheel loads. The depths of the corresponding $0.1p$ pressure bulbs for each shape would be different. Hence, most engineers think roughly "twice B" as the rule of thumb for evaluating the general effects of stress distribution. It is this line of thought that probably led to the following recommendation for soil boring depths:

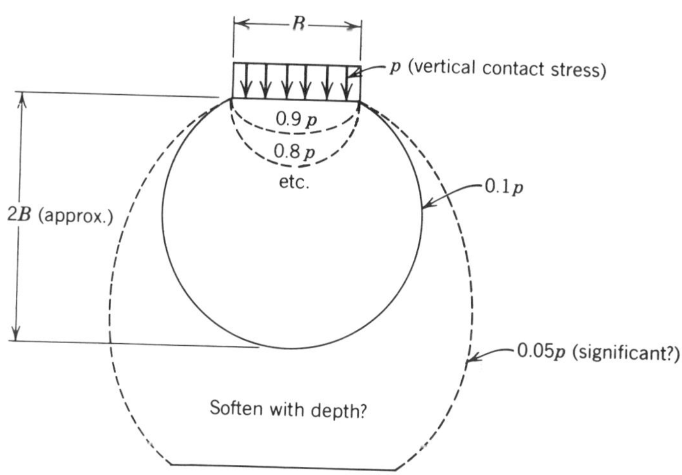

Figure 8.1. *The pressure bulb.*

> Beneath lightly loaded structures with widely spaced columns, the depth of the first boring should not be less than twice the probable width of the largest footing (Peck et al., 1953, p. 139).

Other reasons for using the figure as a guideline rather than a rigid rule are (1) p itself may be unusually large, and hence $0.05p$ may be significant, or (2) the soil below the $0.1p$ pressure bulb may be unusually soft, so that even very small pressures may induce major settlements. Geologically, softening with depth occurs either when the water table has been lowered, or when the entire physiographic region has risen because of tectonic forces. In either case, the region near the surface dries out and forms a crust that is stiffer than the soils below. Such a crust is common in the New Jersey Meadowlands. Conversely, if a number of borings have indicated that bedrock is shallow beneath the proposed structure (i.e., considerably less than $2B$) it would make no sense to follow slavishly the rule of extending borings to the $2B$ depth. "$2B$ or not $2B$, that is the question." (Abject apologies, of course, to W.S.!)

Finally, no soil or rock is homogeneous, Indeed, some have very pronounced layering, as with the varved clays of the Meadowlands, and stress distributions are further affected in a complex way.

To further illustrate the importance of using the pressure bulb to guide one's thinking, I offer a story. When I took my orals for my doctorate, one of the professors on my committee asked what I would consider to be the limiting allowable contact pressures on a typical state highway. Since I had taken a course in pavement design with this professor, the question did not surprise me, and I answered "about 100 psi" by interpolating between a typical automobile (30) and airplane tires (200). The follow-up question was significantly more puzzling:

> Mr. Monahan, are you familiar with a current fashion among the young ladies called "spiked heels"?

I indicated somewhat hesitantly that I was.

> What would you estimate the dimensions of the contact area of a heel?

Answer:

> About ¼-in. square.

Next question:

> Do you have any preference for stature in young ladies, Mr. Monahan, petite versus hefty?

Monahan (getting nervous, now):

> Petite.
>
> What would the contact pressure be for, say, a 100-lb young lady, Mr. Monahan?

(Wielding the chalk, the quick calculation revealed the answer.)

> Uh . . . 1600 psi.

The final query:

> Can you explain, Mr. Monahan, why the state of Oklahoma places a limit of approximately "only" 100 psi on wheel load contact stresses, in view of what you have just calculated?

(Wow! Gulp!) Sure! Fortunately for me, I had had a few years teaching experience by then, which helps one to think better on one's feet, so with a minimum of apparent distress I searched my brain and asserted,

> Why certainly, it has to do with the size of the respective pressure bulb and the manner in which the stresses dissipate with depth!

Since that time, I have related the story in each of my soil mechanics classes, in the belief that in a graphic, humorous, and virtually *unforgettable* way, the concept of pressure bulbs was permanently implanted. One of the funniest exchanges I ever had in class came about when a student offered the "correction" that the contact pressure of the spiked heel should be 800 psi, assuming of course that the girl had two feet, to which I replied, "Only if she hopped like a bunny and in that case, impact load would have to be considered."

The applications of the concept are important and varied. Hopefully, one would now immediately "think pressure bulb" whenever one sees or envisions a loaded area (or areas), rather than make the serious mistake of considering only the contact area stresses (the typical error of structures types specialists). Figure 8.2 shows some applications that come to mind.

Parts (*a*) and (*b*) represent, to approximate scale, the spiked heel and a tire, respectively, bearing on an 8-in. pavement. It can be seen that the bulb of very high stresses of the heel is only about ½-in. in diameter and that the stresses dissipate quickly to nominal levels within the zone of the pavement. By contrast, the stresses associated with the tire extend significantly into the base course and (not shown) into the subgrade (natural soil) below. One can see from these simple considerations that pavement design is an exercise in providing a series of materials (pavement, base courses) of decreasing strengths and stiffnesses, top to bottom, corresponding to the diminishing stresses that will be imposed by the surface loads.

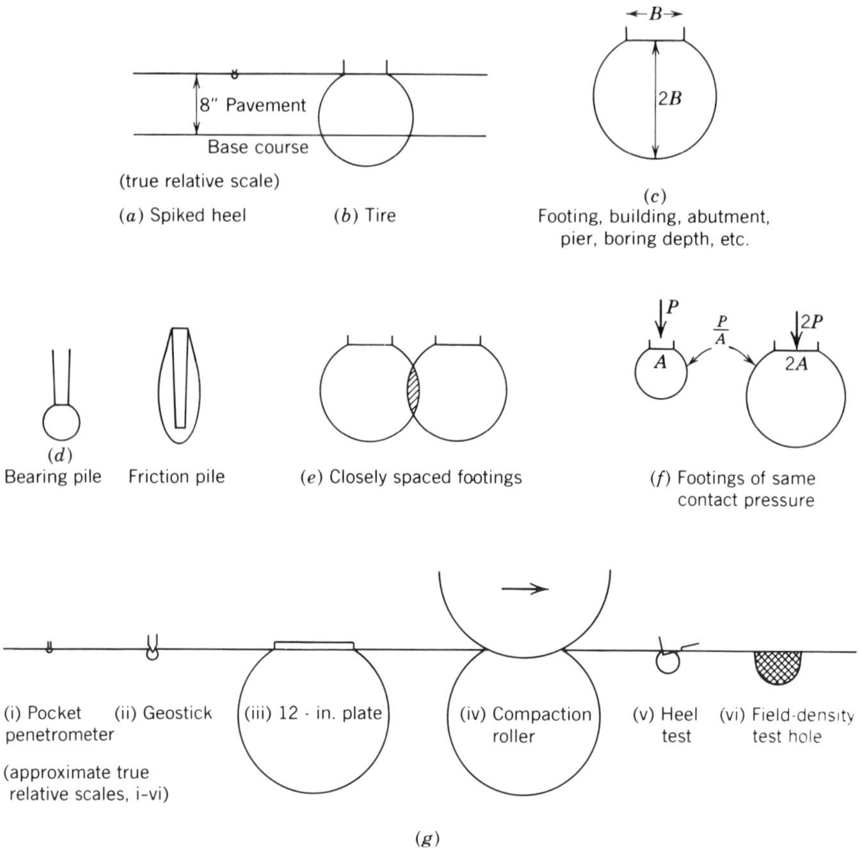

Figure 8.2. *Applications of the pressure bulb concept.*

Figure 8.2*c* is a generalized sketch of any standard foundation element, an isolated footing (one of many supporting a building), an entire building (as for a mat foundation), an abutment, or a pier. The B and $2B$ dimensions are shown merely to suggest a visual device for assessing the zone of influence of significant stresses, a valuable tool for guiding one's judgment in making decisions on the needs for sampling and laboratory testing throughout the duration of any project, from initial conception, to planning the first day's boring program, through the months of modification of one's thinking (a process probably unique to foundation design), to the final selection and detailed design of the specific foundation scheme.

Figure 8.2*d* simply extends the notion to pile foundations. Part (*e*) shows two closely spaced identical footings, illustrating overlapping pressure bulbs. Because of the higher stresses in the zone of overlapping, one would anticipate the probability that the footings would settle unevenly, tilting toward one another.

Figure 8.2*f* provides a definitive answer to the question at the beginning of the chapter. Even though the contact pressures are the same for each footing, consideration of their pressure bulbs leads one to the (now) obvious conclusion that the larger footing will settle more than the smaller, simply because a much larger volume of soil is stressed significantly.

Now we consider specific tools and techniques that relate more directly to fill control operations. These are represented by Figure 8.2*g*. Part (i) depicts a pocket penetrometer, a device that has a spring-loaded plunger or needle about ¼ in. in diameter. One merely pushes the device into the soil and, in seconds, obtains a reading of the unconfined compression strength, sometimes also called the presumptive bearing capacity for clay soils that are to be loaded rapidly. Notice the specific limitations; it applies only to clays and only to those that are to be loaded rapidly. (Rapidly in this case refers to routine construction, measured in months, where no special surcharging and preconsolidation—usually by the installation of sand drains or wicks—is to be done.) But consider the other unstated limitations related to the tiny size of the pressure bulb ($2B$ = ½ in.). Add to this the potential and probable disturbance of the zone of the pressure bulb by construction machinery or by desiccation, and one wonders why the term presumptive is used. The dictionary (Webster's) states, "giving reasonable ground for belief." I would say, "Don't believe it!"

If the use of the device is extended to fills that are other than 100% clay, the presence of a fine piece of gravel (say 3 mm) or even a coarse particle of sand (1–2 mm) could affect the result considerably. A pebble? Forget it! Part (ii) of Figure 8.2*g* shows a larger penetration device, the geostick, available from the Acker Drill Co. It is about 3 ft long, 2 in. in diameter, and has a removable (threaded) cone tip. The handle end is a geologist's pick (and hammer). The tube is hollow and can be used to sample clays (although probably not very stiff or hard clays). The penetration feature of the tool enables one to obtain the (presumptive?) bearing capacity, again in seconds, by simply penetrating the soil vertically with the cone attachment. For soft clays, this is accomplished with the weight of the stick; for stiffer clays with the weight of the user. There is a series of circles on the cone representing increments of penetration. Printed on the barrel is a table relating penetration values and bearing capacities for stick weight, 140, 160, 180, and 200 lb.

The geostick is a pretty handy device, especially since it has multiple uses. Used properly and with good judgment, it is probably superior to the pocket penetrometer, if for no other reason than the larger size of the pressure bulb that it generates. I would also suspect that it can be more readily extended to fill control work on soils other than clays, though not to soils containing large quantities of coarse to medium gravels. The limitations that apply to the pocket penetrometer also apply to the geostick, but logically to a somewhat lesser degree.

Still another type of penetrometer (not shown) is a Torvane device. This is a pocket tool, the main feature of which is a series of blades or vanes. It is inserted by hand into the soil and twisted until the soil fails; the unconfined compression strength is then read directly from a dial. Again, the device is largely limited to clay soils, and judgment must be used in extending its use to others. One advantage of the Torvane is that it can be used to obtain readings from vertical or sloped soil surfaces, which would be problematical with the geostick.

As described in Chapter 7, special criteria such as plate bearing tests are sometimes used in fill control. Because of their expense, these are usually limited to very large earthwork projects. Circular plates are used ranging in diameter from 12 to 30 in., the latter size being a common standard. In some cases, up to three plates of varying size are used to permit more valid extrapolation of the results to the probable performance of the full-size structure (Johnson and Kavanaugh, 1968, p. 347). For pavement design, the 30-in. plate is usually used to determine a subgrade modulus (Yoder, 1959, p. 337). In addition, special tests are devised for specific purposes and applications, such as for the job in South Africa described briefly in Chapter 7. In all cases, the interpretation of load–settlement curves is necessary to obtain whatever answers are sought. Figure 8.2*g*(iii) shows the pressure bulb. Applying the $2B$ criterion, one can see that the pressure bulb can be up to 5 ft deep (for 30-in. plate). Contrast this to the pocket penetrometer! (But also contrast the cost!)

Part (iv) represents a field compaction device and its pressure bulb. No scale is used, for it is intended that the sketch represent anything from an air-operated hand tamper (for ditch work, typically), to a standard roller, to an off-road supercompactor capable of efficient compaction to depths of as much as 10 ft.

Before proceeding to a description of the details of direct field density testing, one last pressure bulb should be mentioned, that of the "tried and true," historic, traditional method of fill control, the famous *heel test* (part (v)). This test is most commonly used in the early phases of a fill control job, in what might be called the getting-things-squared-away phase, before methods have been established that will bring about the desired and specified end-result density. Usually the inspector arrives at the site, and the placement and compaction process is observed. The first thing one does is check the thickness of the loosely placed lift, usually by inserting an opened section of a 6-ft folding rule. Nodding affirmatively (assuming correctness), one then stoops over and obtains a handful of the soil, inspects its texture by rolling the soil through the fingers, looking sort of thoughtful though noncommittal in the process. Then one squeezes the soil firmly in the hand to check for moisture. The compaction rolling is observed, and in some way it is made plain that the number of passes is being recorded, perhaps by holding three fingers up during the third pass. Upon completion of the agreed-upon (or specified) number of passes, it is time for the heel test, which is simply the act of kicking down on the rolled surface several times, accompanied by appropriate facial expressions and head movement (side to side, of course, denoting dissatisfaction). I am being facetious in describing the scenario, but I believe the description is reasonably accurate and that the scene has been enacted many, many times without too much deviation. It's a game that is played over and over.

Even though the heel test appears nowhere in specifications or (until now) in the literature of earthwork construction, the test does have its serious implications. First (and I am here addressing the young inspector), don't ever use the heel test on the surface of a rolled fill that you did not see placed and rolled, especially if the fill is in a confined area such as backfill for a wall or trench, for surface rolling could compact the soil adequately near the surface but have little or no effect on the soil below. In fact, if the soil is at all clayey, a half hour of baking in the sun could dry out the soil and create a very hard, desiccated crust that would give misleadingly good results in the heel test. I know of a case where a fill was approved on the basis of comparable inspection, a footing was built on the fill, and postconstruction settlements caused serious column and floor damage. Consideration of the pressure bulb associated with the heel test will reveal the dangers of placing one's trust in its validity as a sole means of fill control. (Have you ever heard a worse pun?!) The same can be said of more sophisticated tests such as the pocket penetrometer, the geostick, and even 30-in. plate bearing tests of a 10-ft dumped fill. In a word, there is no substitute for observation *augmented* by a carefully conducted field testing program, sometimes utilizing a variety of testing techniques and tools, some for direct measures of density and others (penetrometers) for developing correlations that may be used to complement the control process. As confidence develops in the complementary methods, they may be used as temporary substitutes when time pressures and unforeseen circumstances force periodic suspension of direct measurement.

Part (vi) of Figure 8.2*g* represents the field-density hole, by far the most widespread technique for checking the adequacy of compacted fills.

8.1.2 Density Tests

Field-density determinations can be made in a number of ways, including the sand cone method, the water balloon method, the jacked sample method, the chunk sample method, and through the use of nuclear devices. All but the last are destructive tests in that they require obtaining a sample of the soil, weighing it, getting its moisture content, and determining its volume.

THE SAND CONE METHOD Figure 8.3 is a photograph of a field-density test in progress using the sand cone method. End-result field densities are designated in specifications as some percentage of a laboratory determined maximum dry density: the target value. As explained in Chapter 3, the dry density is computed from three variables: the weight of the soil in the hole, W; the volume of the hole, V; and the moisture content of the soil, w.

Figure 8.3. *Sand cone field density test.*

To conduct the field test, one first selects the spot and elevation to be tested (more on this later), and then prepares a fresh surface of the compacted fill for testing. As a minimum, this usually involves scraping off the top few inches of the rolled fill to eliminate the artificial effects of shrinkage by desiccation. (This may or may not be covered by the specifications; perhaps it should be.) Typically this might require tedious hand work with a shovel and other tools. In the photograph, the surface was prepared by passes of a bulldozer blade, at my direction. A level, smooth spot was chosen (obviously between the tracks of the treads), and the plate set down in a firm, flush position. (With extensive amounts of gravel and cobbles present, as was the case here, finding a spot to test is not always as simple as saying it. Even then, the content of the hole may invalidate the test, as we will see.) The density hole was dug in a roughly hemispherical shape, and the contents were placed in the can for purposes of determining the soil weight W and its moisture content w.

Before doing so, however, one must make a qualitative judgment on the acceptability of the contents of the hole in terms of texture variation compared to the laboratory test borrow. Remembering that the essence of valid laboratory testing is to *simulate field conditions*, the soil in the hole must reasonably conform in texture with that which was tested in the laboratory. To define reasonably, what I am saying is that you should not have a 3-in. cobble in that can, such as the one shown in the left foreground. If you blindly accept such a "dig," it is predictable that the computation for the density will turn out to be about 160 pcf, the density of solid rock being about 170 pcf. (Moreover, if you do accept the test, and report its results to your project engineer, he will wonder about your wrapping.) In Chapter 3, it was noted that the laboratory test requires the exclusion of larger particles, depending on the size of the compaction mold used (see Section 3.4, ASTM Compaction Requirements), and that the percentage of the excluded material is recorded. Assuming that the soil from the test hole reasonably conforms to the texture of the borrow, what do we do about the inevitable difference between the percent (+ *n*) excluded in the laboratory test compared to the corresponding percent in the test hole? The solution: an empirical correction. (The formula that was used is proprietary.)

Returning to the field density test, it remains to explain the techniques used for determining *W*, *w*, and *V*. The weight *W* is determined straightforwardly by weighing the can of soil and subtracting the tare weight of the can. Typically, the field inspector will work off the tailgate of a vehicle and will be supplied with a standard triple beam balance of rugged design and suitable capacity and sensitivity. To the young inspector: Learn how to use it. Don't forget the hanger weights.

The moisture content *w* may be determined by "cooking" or by the use of a "speedy moisture" device. The cooking technique is the same used in the laboratory, except that one would use a Coleman camp stove or similar instead of the usual electric oven of the laboratory. For convenience and to improve accuracy, the entire can of soil can be cooked to constant weight, and the moisture determined by before/after weightings. The speedy moisture technique involves the use of a commercially available device utilizing a measured powdered chemical that is mixed with the moist soil in a chamber, producing a pressure. A gauge is attached that is calibrated to read moisture content directly.

The volume *V* of the test hole is determined by filling the hole with a material of known unit weight, in this case Ottawa sand. This sand is commercially available and usually sold in 50-lb bags. It is a sand of remarkably consistent texture, and when poured into a hole or other container in a standard fashion will assume a density of about 99 pcf. If you look carefully, you will see that the base of the cone has a lip that conforms to the hole in the plate. To conduct the test do the following:

1. Weigh the sand and jar, with cap (in advance).
2. Remove the cap and attach the cone to the jar.
3. Check that the valve on the cone is closed.
4. Invert the cone and place it on the plate rim. (Be sure that the rim is free of soil particles.)
5. Open the valve.
6. Close the valve when flow has ceased. (*Note:* This is, for some reason, fairly easy to forget.)

7. Disconnect the cone and reweigh the jar and cap (residue and jar).
8. Subtract (7) from (1) to obtain the weight of sand used to fill the hole and the cone.

Steps 1–8 will yield the field data needed to determine the volume of the hole, V. All other required data are normally determined in the laboratory. Two items are included: the weight of the sand necessary to fill the cone and plate (this, subtracted from (8), yields the weight of sand in the hole), and the unit weight of the sand. The procedures for determining these items are simple, standard laboratory procedures and would probably be performed by a technician. While there are many details of the field and laboratory procedures, just remember what you are after and that you are simply filling a hole of unknown volume V with a material whose unit weight you have determined. Thus, if the sand used to fill the hole weighs 3.30 lb and the density of the sand is determined to be (in fact) 99.0 pcf, the volume V is 0.0333 ft^3.

Schroeder (1980, p. 139) presents a complete set of data for the sand cone method. This test, and most others noted in this book have ASTM designations, and are described fully in ASTM publications.

When the test is completed in the field, it is a good idea to retrieve most of the spilled sand for reuse by shoveling it into a clean canvas bag, not so much because it is expensive, but rather to avoid running out, thus necessitating delays and repeated laboratory calibration work on a "new" sand. When you do this, avoid contaminating the sand with soil from the site.

Finally, the weighing and moisture content determination of the soil from the test hole should be done immediately after the sand weight is ascertained, to avoid errors through drying. This would be especially important on a hot, sunny day.

THE WATER BALLOON METHOD A more recent method for determining the volume of a density hole is through the use of a rubber balloon apparatus, where water is used to fill the hole. A thin latex rubber balloon within the hole contains the water and stretches to conform to the shape of the hole. Interestingly, in the earliest years of field-density testing, a board with a hole was placed over the test hole, a balloon was stuffed into the hole by hand, and water was poured into the balloon. The Ottawa sand cone method apparently was developed because of the inherent inaccuracy and sloppiness of the original water balloon method. The new water balloon method utilizes a device consisting of a graduated glass cylinder housed in an aluminum protective guard, with appurtenances allowing for quick, neat, and accurate determination of the volume of the test hole. The device is called Volumeasure and is available from Soiltest, Inc., Evanston, IL. Other similar devices, such as the Washington densometer, are used.

The devices were not available when I was doing fill control inspection, but I understand through talking with those who have used them that they are quicker than the sand cone method, each test taking about 20 minutes. The sand cone method I would judge takes perhaps twice as long but these estimates would be contingent upon the texture of the fill, particularly as it may affect the preparation (scraping) and digging of the hole and the unpredictable interference of gravel and cobbles. The water balloon method would seem to be inherently more accurate because of its use of water, which, of course, has an invariant density (for all practical purposes) and requires no calibration. Water is also cheaper than Ottawa sand.

JACKED SAMPLE METHOD Having done site investigations involving test pit inspection and evaluation, I developed a means of extracting relatively undisturbed samples by jacking a section of scrap shelby tube (a hollow metal tube), about 6 in. long, into the wall of the test pit, using a 7-ton aluminum jack. Figures 8.4 and 8.5 illustrate the method. After jacking the tube into the wall, extraction is accomplished by digging the filled tube out of the wall with a geologist's pick or sturdy claw hammer. The samples thus obtained can be used for testing and analyses as with any undisturbed samples, including textural classifications (sieve and Atterberg limits), unconfined or triaxial compression testing, consolidation testing, and density–relative density determinations. This capability, in my opinion, more than compensates for the lack of blow count data that would routinely be available through conventional boring operations with a drilling rig.

For those who might use the technique for general foundation testing and analysis, the effects of anisotropy may be significant, inasmuch as the samples obtained will be approximately 90° out of the normal vertical orientation, and strength–deformation test results may be seriously affected accordingly. Textural and density test results will, of course, be unaffected. When deemed necessary, or when convenient, samples can also be obtained by jacking in a vertical direction, in which case, however, you will need something to jack against, like the bucket of a backhoe. The disadvantage here is that you'll need a backhoe and a backhoe operator to assist. Horizontal jacking can be done alone.

The jacking method described has been used only in test pit sampling of natural soils. Whether it can be easily extended to sampling compacted fills I am not certain, but I would expect that the texture of the fill and the energy of compaction would need to be considered. For example, an essentially granular TALB, containing significant quantities of coarse to medium gravel, compacted to a high percentage of modified Proctor, would be very difficult to sample by jacking. It would also, I might add, be very difficult to dig in (for the sand cone or water balloon method). These would be the blister densities that I referred to earlier.

CHUNK SAMPLE METHOD As surprising as it may first seem, the very best type of sample one can use for general laboratory testing is a plain old chunk of the soil, assuming that reasonable care is taken in obtaining the chunk. A typical method of obtaining such a sample is to isolate a knob of the soil by carefully

Figure 8.4. *Jack sample method.*

Figure 8.5. *Extracting a jacked sample from the wall of a test pit. (Whadd'ya mean, it's the wrong lot?)*

digging a circular trench; the chunk is then obtained by using a sharp spade to cut and lift out the chunk. Another method would be simply to tear or pull a sample carefully from the wall of a test pit, avoiding a severe mashing or shearing action. The reason that such samples may be regarded as best is that the act of sampling does not significantly damage the sample. All other types of conventional sampling (e.g., standard penetration and shelby tube) require driving or pushing a sampler into the soil, producing disturbance by shock and/or side shear between the sampling device and the soil.

The moisture condition of a chunk sample can be preserved by wrapping the sample with a clinging type of clear food wrap. If necessary (for long storage), the sample can be further encased by smearing it with melted wax.

The principal difficulty of strength–deformation testing of chunk samples is that one must trim the sample into some prismatic form, usually a cylinder. For density determinations, however, this is not necessary if one can devise a method for measuring the irregular volume of the chunk. One suggested method is to fashion a chunk-volume determinator. A large can with an overflow weephole and attached small-diameter graduated cylinder (or pipette) would suffice. One would fill the can with water to the level of the weephole, empty the pipette, and then carefully submerge the wrapped chunk sample into the can. The water displaced into the attached pipette would be a measure of the volume. The weight of the sample and its moisture content could then be determined in the usual manner, and the dry density computed. In doing this test, it would be important to wrap the chunk very carefully to ensure that there are no air bubbles of significant size trapped in or between the wrapping folds.

THE NUCLEAR METHOD Density determinations by nuclear methods might also be considered. While I have no first-hand experience with the use of the device in the field, I understand that the apparatus emits gamma radiation into the soil and that the measured rebound is calibrated to the density of the soil. In a similar way, the moisture content is determined by using a high-energy neutron source. As might be expected, these devices are very expensive. When calibrated and working properly, their major advantages are instant answers and speed of testing, thus enabling the performance of many tests in a short period of time. I understand that for highway work, a unit can be mounted on a jeep and test readings can be taken virtually on the run.

Their high cost limits their use, sensibly, to very large jobs or to organizations that routinely do major earthwork jobs, such as earth dams or highways. For smaller companies, which would probably not use the device day in and day out, I understand that there is a problem with maintaining the equipment in reliable working order; that is, the equipment tends to get out of calibration with disuse. Another problem is the bureaucratic involvement with the nuclear regulatory officials in Washington. My perception of these disadvantages stems from my only involvement with the device, which occurred about 30 years ago. A friend and colleague from a local geotechnical consulting firm gave the device (sans nuclear element) to the college for instructional purposes. That's when the letters from the AEC (Atomic Energy Commission) started coming. It turned out that merely owning the *container* that once housed the element required conformance with certain procedures.

Whether or not the newer models of the nuclear density gauge have similar calibration problems, I cannot say for certain. I would expect, however, that regulatory involvement is, if anything, more intense. My suggestion to those who would consider the device: Check it out first.

8.1.3 The Testing Program

After a review of the important features of the pressure bulb and the various ways of performing field-density tests, the following questions arise: Which test method(s) should we use? How much field testing is enough? I have no pat answers, except to say that a lot of good judgment and experience will be needed to provide good answers.

The methods one uses will depend, among other things, on the size of the job; the texture and homogeneity (or lack) of the borrow; the experience, expertise, and preferences of the controlling agency; and (to be blunt) the intelligence, experience, and even the personality of the inspector. Hopefully, the choice of methods and how to employ them will be aided by descriptions in the preceding section and elsewhere in this book. The second question, relating to decisions regarding how much testing to specify for a given job, is a much more complex issue. Included among the considerations, I would recommend the following:

1. Cost
2. Consequences of distress or failure
3. Environmental factors
4. Degree of verification: expected versus warranted

Before getting into some commentary on each of these items, it will help to set a perspective by introducing the notion of a *sampling ratio.* First, we can assert that it is clearly not possible to test *all* of the fill, so we have trapped the answer: somewhere between zero and all, and undoubtedly closer to zero. (That's not much help, but it's a start!)

In the fill job cited earlier for a major earth dike system, Hendron and Holish (1980) report one density test for each 2000 yards of fill, or approximately 1500 tests for the 3 million yards of dike. Assuming 6-in.-diameter holes approximately of hemispherical shape, the total amount of soil tested was 2.42 yards, yielding a sampling ratio of 0.809×10^{-6} or, in round numbers, $1/10^6$. Utilizing the pressure bulb concepts presented in Section 8.1.1, similar analyses could be made for any project, whether for soil sampling, destructive testing (field densities, as above), or nondestructive testing (plate bearing, for example).

I am not aware to what extent this kind of analysis has been done in professional practice, inasmuch as my career has been largely academic. What professional experience I have has been largely on small jobs where the number of borings, amount of testing, etc., is largely and more easily determined by the common sense and experience of senior engineers. It may be that the process of pursuing the elusive answer of "optimum sampling ratio" (if indeed one exists) is extensive in practice, but if so, the investigations and results do not appear to have been published yet, at least as far as my literature review has revealed. Such a "techonomic" analysis would be very interesting and valuable.

A number of reasons might explain the dearth of information in this area, the first of which is that none of those who are knowledgeable enough and privy to sufficient data have thought of the idea of publishing an article or paper on the subject. Or it may be that they simply don't have the time. Clearly, it would be only those people in full-time professional practice over an extended period who would possess both the knowledgeability and data to produce such a work. Unfortunately, it is only the academics who have the luxury (i.e., time) to produce scholarly papers. Indeed, it is (or should be), next to teaching, their responsibility to do so. But many lack *any* experience (or interest), and almost all lack the data. Another reason for the lack of literature on this important topic might be that the data available within a professional organization are considered proprietary. Since it takes considerable time to collect, compile, and analyze such data, it is understandable that one would not wish to share them with competitors.

How may this void be filled? I see two possibilities. First, the data from public jobs, that is, those done by government organizations and therefore using tax revenues, should be available for the asking. The second avenue of approach would be an appeal by an appropriate organization for the voluntary contribution of the needed raw data. I would suggest that the Geotechnical Division of the American Society of Civil Engineers consider this solicitation. The compilation and analytical work could be done by academics with strong interests (and some experience) in the area, including professors and the graduate students whom they supervise on research projects and thesis work. To enhance the validity of the work, communication could be established between the contributing firm and the academics, including the participation of professionals on student (thesis) advisory committees. From recent first-hand experience, I can say that there is strong interest and movement toward greatly increased cooperation between industry and universities. This, it seems to me, would be an ideal way to fill the techonomic void that seems to exist in geotechnical engineering, while at the same time serving the broader purpose of industry–university interaction. It would appear to me that both would benefit. I further believe that this would be a legitimate activity for academics in that the work would be serving the needs of industry rather than competing, inasmuch as—if I am correct—industry does not have the time to pursue such applied research. Finally, I also believe that the economics of engineering practice is a badly neglected area of study and understanding among academics. From a recent conversation I had with an experienced consultant, I gather that there exists considerable lack of awareness of the effects of economics on design decisions among practicing engineers. Thus, the approach I suggest may have beneficial effects on both ends.

COST FACTORS Sowers (1979, p. 291) has reported that:

> the cost of an adequate investigation (including laboratory testing and geotechnical engineering) has been found to be from 0.05 to 0.2% of the total cost of the entire structure. For critical structures or unusual site conditions, the percentage is somewhat higher, from 0.5 to 1%.

While I have not met or communicated with Professor Sowers, I was interested to note the dichotomy he makes between costs of geotechnical work for structures and critical structures, an approach similar to that which I adopted independently in devising my concept of a design flowchart (see Figure 1.2), where I refer to cheap jobs and expensive jobs. As described earlier, the former relates to jobs where the total cost is low and the consequences of distress are not severe, allowing one to design on the basis of the use of inexpensively obtained index properties, published charts, and experience. Expensive jobs were for structures (Sower's critical structures) where the consequences of distress or failure are of acute concern, thus sensibly prompting decisions for more expensive sampling and testing to determine engineering properties necessary to design a more foolproof foundation.

Buried among Sower's cost estimates is the fraction that might be allotted to fill control costs. So one approach to deciding how much density testing to do (i.e., choosing a sampling ratio) would be to base the decision, at least partly or initially, on budgetary considerations and then see how the answer conforms to technical judgments. This approach is, of course, nothing new to business people who often start out by asking, "What can we spend?" then working backward to flesh out the line-item details.

It is interesting to note that Professor Sowers is designated in his book as Regents Professor of Civil Engineering, Georgia Institute of Technology; Senior Vice President and Consultant, Law Engineering Testing Company. In his preface, Professor Sowers also notes that there is "more than a century of geotechnical experience in the Sowers family, spanning three generations." One can see that Professor Sowers is one of those rare individuals who somehow finds the time to both do the work and write about it, hence the valuable data he graciously shares with us on costs.

CONSEQUENCES OF DISTRESS In considering the multitude of decisions regarding costs and the concomitant details of foundation work for a structure, one probably subconsciously thinks about what would happen if the structure cracks, moves, or even fails completely. From a scientific or technological point of view, however, it would be much more beneficial to the design process if one *consciously* considers such factors, since the impacts can range from the trivial to the catastrophic in terms of costs measured in many ways. Tangible losses could include property and perhaps even lives, and intangible losses those of reputation and peace of mind. (If this seems overly dramatic, I remember that an engineer committed suicide after the collapse of the mountain wall of the Vaiont Reservoir in Italy, which resulted in the death of 2000 people in the early 1960s.)

Details of these considerations could include the following: What will be the use of the facility? As a mundane example of a compacted fill, consider a fill needed for a parking area for light trucks and automobiles at the rear of a warehouse, a job on which I worked many years ago. Our recommendations were to compact to 95% standard Proctor, install an asphalt (flexible) pavement, and grade for surface runoff away from the structure. We further stated that surface patching would be needed in the future.

In this example of a cheap job, without going into detailed cost studies, the project engineer decided that the cost of modified Proctor compaction and a high-quality pavement was not warranted and that considerable settlement was tolerable. For that reason, the flexible pavement was recommended, one that would conform with the settlements as they occurred. If too much local settlement occurred, resulting in puddle formation, inexpensive patching could rectify the problem. As I recall, the amount of field density testing was zero, the control being accomplished by observation on site.

If the fill were to support a slab on grade, covered by a terrazzo tile for the reception area of an office facility, the recommendations for compaction, testing, and field control would have been significantly more stringent and, consequently, more costly.

ENVIRONMENTAL FACTORS What about nuclear power plants? I am *not* suggesting that nuclear power plants be built on compacted fill! I *am* suggesting you read a book that deals with matters of risk, safety, and consequences in a readable and interesting way: *Of Acceptable Risks*, by William W. Lowrance, Harvard University. The book is short (180 pages), scientific, but with almost no heavy mathematical treatment, and, above all, timely. I believe it will be of interest not only to those involved in construction but also to anyone who is concerned about the burgeoning moral and technological complexities of our society.

With respect to the more prosaic aspects of consequences of distress and how they affect the testing program for fills, one must not fail to consider the applicable laws, building codes, and permits dealing with the construction, both from a safety standpoint (OSHA) and those dealing with potential environmental consequences, the latter becoming more and more prevalent in all areas of construction. Indeed, while I am not qualified to go into the subject, *hazardous* landfills have become a major national concern. Having lived and worked in urban New Jersey most of my life, I am more than normally sensitive to the problem, professionally and otherwise. In a survey conducted by the U.S. Environmental Protection Agency (August 1983), New Jersey was identified as having the largest number of sites containing hazardous waste materials (65 of 406). As a specific example of hazardous fills, emissions of radioactive radon gas from the soil surrounding several homes in a neighboring town have been detected recently. It seems that many years ago a company that made watches with luminous dials (radium) was active in the area. Speculation is that they disposed of their wastes on site, but that fill was later obtained from the abandoned site for grading of other residential lots. And, of course, everyone has heard about Love Canal. I expect that within 10 years a book will appear in this series entitled *Avoiding Construction of and on Hazardous Fills.*

In terms of the testing program, I can only speculate that there will develop a need, indeed, enforced requirements, for more tests for toxic and hazardous substances, particularly in the more industrialized, urban areas and within haul distances of those areas, Then again, Times Beach, Missouri, no metropolis, was rendered uninhabitable by the spraying of dioxin-laced oil, done for the purpose of controlling road dust. (Dioxin is a deadly poison.) Since the publication of the 1st edition of this book, a law has been passed in New Jersey called The Environmental Clean-up and Responsibility Act (ECRA). This legislation led to the generation of voluminous regulations requiring, in many cases, very large expenditures of money and effort to clean up properties that were found to be contaminated. Some of these regulations were, in my opinion, draconian. In one case, for example, a successful restaurateur purchased an adjoining piece of property to create additional parking, only to discover that the site had been severely contaminated by gasoline from leaking buried tanks from a service station that had once occupied the site. Notwithstanding the fact that the new owner was in no way responsible for the contamination, he was ordered by "the authorities" to clean it up by a certain date, after which enormous daily fines would be imposed if he did not comply. As might be expected, there have been constant and numerous objections to many parts of this act by business people and developers. With the recent downturn of the economy (in New Jersey and nationally), the state has relented and substantially modified the "more draconian" aspects of the regulations, much to the relief of the business community.

This is an example of how environmental concerns are greatly affecting many aspects of our daily lives, including the "small world" of geotechnique and construction of fills. Chapter 9, Septic Systems, will further illustrate this important new development.

VERIFICATION Just as it is not possible to test all of the fill, neither is it possible to provide *absolute* verification of the compaction specifications, and clients, lawyers, and regulatory agencies should understand this. Other factors that can complicate the problem of verification, many having little to do with technical considerations, but seriously affecting the testing program and its cost. For example, it is normally the project engineer on a job who signs his or her name to the report or letter that certifies compliance with specifications. Typically, at least in my experience with small jobs, the project engineer supervises field inspection work from the office, with only occasional visits to the site. As described earlier, present earthwork construction practice is such that the field inspector is likely to be inexperienced. Hopefully, with the advent of civil engineering technology programs, and perhaps the future development of other technical support people, inspection work will be greatly improved by the creation of a group of people, whatever their titles, who will develop the skills and experience to provide the needed field control. I further suggest that such people will have to be compensated at higher levels as their skills and judgment improve with education and experience. The apparent costs of construction testing programs (i.e., fill control) will increase, but it is my belief that the actual costs will go down in the long run by virtue of the greater cost reductions related to reduced postconstruction damage and the attendant lawsuits. I believe the current practice of "kicking young inspectors upstairs" about the time they learn to perform well is self-defeating.

In any event, the degree of certification that will pertain to any job will be a significant factor in planning the testing program. There are, perhaps, three levels of certification that I can perceive:

1. That which is governed by the expectations of the client.
2. That which is legally required.
3. That which is dictated by the engineer's own standards of performance and reputation.

The client may be a company that has its own engineering staff, in which case communication can be established to ensure that there is an understanding on the part of both parties of what is intended and expected. If the client does not have engineering advice (internally), it is likely that he will depend exclusively on your professional reputation to do what is needed. In this case, I suggest that it still makes sense to communicate your intentions *and what it will cost* (at least in approximate terms) so that there will be no future shocks or major surprises. Continuous communication, as inevitable changes occur, will also be beneficial to all concerned parties—trite perhaps, but no less true.

On some jobs, certification of construction operations like filling is legally required. Obviously, in such a case, the extent of field testing, and thus its cost, will undoubtedly increase, Case Study 2, Section 4.2.1, was such a job. Not only did the specifications call for supervision and certification, but it developed that the source of these requirements (particularly the certification) was a state regulating agency in Newark, NJ, to which I was required to send a letter certifying the fill. If you reread Case Study 2, you will see that this created a rather sticky situation. (To this day, I do not understand how a regulating agency in

Newark got involved with a small construction job 40 miles away. Numerous other similar jobs in New Jersey that I have done did not have such a requirement. I didn't ask.)

Finally, if there are no legal requirements for certification, and the client is of the do-what-you-need-to persuasion, the degree to which you certify the fill is up to what might be called "company policy." And here it can get a little delicate, especially in view of our litigious society, a subject I touched on briefly earlier. For now we come to what lawyers call exculpatory clauses. These are statements that are inserted in reports, contracts, and plans that are designed to evade responsibility for anything that subsequently goes wrong. In legal parlance, they are also called hold harmless clauses.

I do not intend here to get into the legalities of the language of contracts, for I have little expertise or experience in that area, but I do wish to make some comments pertaining to the language contained in correspondence with clients, usually in the form of a cover letter accompanying a full report. The phrases that have always disturbed me are those that say, in effect, "We are not responsible for anything in this report." Call it exculpatory, call it hold harmless, I call it avoiding responsibility. Of course, such statements never appear in the plain language I give as an example. The wording is more subtle, typically very brief, and usually occurs in the very last paragraph (I suspect, so that the reader won't be as likely to notice). I have seen similar types of disclaimers stamped on drawings, very commonly on boring logs and soil profiles. My reaction when I see such phrases is, "What are we being paid for?" (as an engineering firm) or, with some sarcasm, "Does this mean we don't get a bill?" (from the drilling contractor).

I hasten to make an important distinction between generalized, all encompassing exculpatory statements and very legitimate statements defining limitations. An example of the latter is afforded by the following paragraphs, which were standard inclusions in all reports of the firm of Woodward-Clyde-Sherard and Associates:

> Limitations—The recommendations made in this report are based on the assumption that the soil conditions do not deviate appreciably from those disclosed in the borings . . . If any variations or undesirable conditions are encountered during construction, the Soil Engineer should be notified so that supplemental recommendations can be made.
>
> This report is issued with the understanding that it is the responsibility of the Owner or his representative to ensure that the information and recommendations contained herein are called to the attention of the Designers and incorporated into the plans, and that the necessary steps are taken to see that the Contractor and the Subcontractors carry out such recommendations in the field.

Notice the reasonableness of these limitations. Notice also that there is an implied inference that "we are confident in, and proud of, our professional competence and the correctness of our recommendations." In my view, that is a very important part of the successful practice of professional engineering: development of a feeling of confidence on the part of the client that the work for which he is paying is of the highest possible quality. It seems to me that one destroys this image by incongruously adding exculpatory statements of the kind described.

It may also be helpful to point out that exculpatory clauses do not always hold up in court. Richards (1976, p. 34) describes a case where the State of New York was "held liable for damages . . . despite exculpatory clauses in the contract." This was a situation where subsurface conditions were "vastly different than originally forecast on the plans" (supplied by the state). *Civil Engineering*, the official magazine of ASCE, advises that "an owner cannot be assured of absolute protection by exculpatory clauses" (May 1975).

With respect to a final report to a client, then, it would seem to me that some verification of the fill compaction should be supplied, but with appropriate and reasonable limitations included, with perhaps specific reference to the sampling ratio and its relationship to the degree of verification that is possible. Carefully chosen language will protect the company both legally and with respect to its professionalism in dealings with clients.

Finally, one should recognize that, whatever the testing program, its cost, and its sampling ratio, it is the quality of the field testing work that is the most important determinant of whether or not the assessment of the fill is correct. It's like that most infallible system of computer programming, GIGO—Garbage In, Garbage Out. Better a bright, observant, experienced person with a pocket penetrometer, than a costly testing program of sophisticated tests conducted by an inspector with little sense of responsibility.

8.2 THE COMPLEAT FIELD INSPECTOR

Before proceeding with a detailed description of field activities of the fill inspector, which I shall do with a case history approach, I would like to offer some commentary relating to equipment that a person who expects to be active in construction field inspection work can use as a check list in preparing for a field assignment. It is also intended and suggested that an employer (such as a geotechnical consulting firm) consider maintaining supplies of certain items so that any employee who is given a field assignment will be able to sign out the needed gear, with assurance that it will be complete and in working order. As I envision the process, it would be the responsibility of a staff person in the firm, perhaps a laboratory technician, to maintain the equipment in working order, conduct necessary inventories, and order replacement supplies and equipment as needed. With such procedures, the chance of a person discovering an important or vital piece of equipment missing upon reaching the site will be minimized.

Because I have been involved and interested in field work for all of my career, I have compiled notes on many aspects of field work, including general reconnaissance work, boring inspection, test pit inspection and sampling, load testing, and, of course field-density testing. Rather than cull the information pertaining only to fill control work, I include it all in the hope that the information will be more broadly useful. Certainly, no field person will be so highly specialized as to be assigned only to fill inspection work, so a more general treatment makes sense.

First, the list. Any item followed by an asterisk is explained more fully, for clarification, after the list. Items without asterisks are either self-explanatory or are described elsewhere in the book.

A. Personal Gear

1. Clothing of appropriate capacity*
2. Foul-weather gear:
   (a) Rain: slicker, rainhat (hard hat), boots
   (b) Cold: handwarmer, insulated boots, special socks, suspenders, wool watch cap or ski mask, writing gloves, ski mittens, lip balm, facial cream*

B. General Equipment

1. Distance measurements:
   (a) Knowledge of pace length*
   (b) Metallic tape (50 ft)
   (c) Set of chaining pins*
   (d) Stanley steel tape (retractable)
   (e) 6-in. pocket ruler (zero end = physical end)
2. Elevation measurements*:
   (a) Folding rule (6 ft)
   (b) Locke level*
   (c) Level rod (or facsimile)
3. Direction measurements—compass*
4. Record keeping*:
   (a) Pocket notebook (plus protective plastic envelope)
   (b) Pen, pencil sharpener (*no eraser*)*
   (c) Camera, flash bulbs, film (fresh batteries)
   (d) Speed message forms
   (e) Pocket cassette recorder (fresh batteries)
5. Slope and plumb*:
   (a) Carpenter's level

   (b) Plumb bob
6. Tool bag

C. *Geotechnical*
1. Penetrometer(s)—pocket*
2. Torvane*
3. Geostick*
4. Sample bags (small, heavy plastic)*
5. Sample bags (large burlap)*

D. *Miscellaneous*
1. Hand sledge
2. Stakes
3. Keel (crayon)
4. Large shovel*
5. Garden spade*
6. Geologist's pick*
7. Soft rock hammer

E. *Special (as appropriate)*
1. Test pit sampling*:
   (a) Shelby tube sections (approximately 6 in.)
   (b) Lightweight hydraulic jack (7 tons)
   (c) Jacking kit (wood blocks, base, plate, tee)
2. Field-density testing*:
   (a) Scale (rugged, triple beam, counterweights)
   (b) Oven (or other device for moisture determination)
   (c) Sand cone apparatus (or equivalent, for hole volume)
   (d) Sieves (full size #4, full set of miniatures)*
3. Load testing*:
   (a) Engineer's level
   (b) Aluminum cubes (approximately ½ in.)
   (c) Flashlight
   (d) Hand telescope or binoculars

CLARIFYING COMMENTS

A.1. By clothing of appropriate capacity I mean that which will accommodate all tools, devices, and record-keeping equipment that can, or should be, carried on the person. Thus, clothing with many and large pockets should be worn. For example, many people would not carry a notebook and pencil on their person, opting rather to leave writing materials in, say, the construction trailer. The trouble with that procedure is the strong likelihood of forgetting to write down some important observation that may have been seen down at the bottom of an excavation, for example, or at some other location remote from the trailer. Sometimes, of course, the observation will include measurements as well, so the pad and pencil, *and* the measuring device(s) should be on the person. There is, of course, a limit to what one can comfortably and reasonably carry, and circumstances will usually dictate sensible choices. But certainly a pad, pencil, 6-in. ruler (maybe a Stanley tape), and possibly an inexpensive small camera, worn on the belt, should be carried at all times. Other equipment on the list that were deemed suitable in advance for the particular job would be in the trailer or in your car for use when needed. A camera is strongly recommended, whether carried on the person or not, in that very important events can be recorded more or less infallibly. Also, much time can be saved by eliminating sketching and writing that otherwise would have to be employed. Sketches and notes, moreover, are considerably more subjective!

A.2.b. Cold weather gear is a necessity, not only for personal comfort (and perhaps frostbite prevention in *really* cold weather), but also to enable one to perform the field work, much of which involves a great deal of writing. Pile inspection is probably the best example in that driving records of the hammer blows must be recorded per foot of driven pile, and more often than that when the pile reaches its required degree of resistance (called "fetch-up"). Writing with nearly numb hands is no fun, and it gets to be virtually impossible to do so legibly. Hence, the handwarmer suggestion for the cold weather inspection job. There are pocket handwarmers available that utilize lighter fluid. You merely light the wick in the morning for a few minutes, blow it out, put it in a protective pouch, and (to my amazement) it stays hot all day, even in the coldest weather. There are also solid fuel types available.

The other cold weather gear cited is intended to keep the other extremities warm, or at least reasonably so. Outdoors experts recommend silk socks under heavy wool socks, the effect being a trapped layer of warm air for additional insulation. If you are especially sensitive to cold, you might even consider electrically heated socks, particularly if you have a long-term cold weather assignment and the nature of the job requires more standing than moving, such as pile inspection. I have found the difference between standing and moving jobs to be dramatic in terms of keeping the feet warm. Especially avoid tight footwear. A boot one size larger than one's street shoes may be helpful to accommodate the heavier socks.

While I have never confirmed it personally, I recently was told by a construction worker that suspenders, rather than a belt, are more conducive to warmth. This makes sense, since the entrapped warm air would serve as added insulation, as with the layered socks. Layered, loose clothing is, in general, the better choice.

The watch cap is important for two reasons. First, it directly protects the ears, and second it helps keep your feet warm. Say it again, you say! Listen: If the body is regarded as a heat sink, when heat is lost, where will the painful sensations be felt first and most intensely? The scientific answer, at least from a heat loss standpoint, is where the area–volume ratio is highest. And where is that? The ears, the fingers, and–of course–the toes. So wearing a hat keeps your feet warmer. (Dr. Gabe Mirkin, fitness expert for WCBS radio, talking about running in cold weather, reported that at 20°F, 40% of the body's heat loss would be through an uncovered head; at 5°F, 70%.)

The ski mask may be advisable for extremely cold windy weather, when frostbite is a distinct possibility. For extended cold weather assignments, a beard makes good sense. Women and clean-shaven men can use facial cream.

Writing gloves refers to a pair of gloves that will provide some protection but still allow you to write legibly. These might be helpful on jobs where almost continuous writing is required, as with driving logs. Leather ski mittens could be worn over the finger gloves when writing is not required. (When your mom sent you out with mittens, she knew all about Monahan's area–volume theory.)

B.1.a. A field person should determine his or her pace length. A simple way to do so would be to lay out a measured distance of 100 ft with a metallic tape, and simply count the number of paces to cover that distance. Do not exaggerate the individual steps, as football referees do, but rather walk at a comfortable gait. It is not only a matter of comfort, but also one of accuracy, in that you'll often need to determine fairly long distances in the field and striding unnaturally will be both tiring and inaccurate.

B.1.c. When a distance measurement should be more accurate than pacing would seem to permit, a 50-ft metallic tape and a set of chaining pins may be used. The chaining pins permit a person to tape a distance alone, since metallic tapes have a rectangular metal loop at the zero end. You proceed as follows: Pin the loop into the ground, go forward 50 ft, and mark the point. Then pull the rear pin forward as you would a fishing line. Chaining pin sets contain two rings that may be attached to one's belt (or suspenders!). One starts with all pins on one ring. As the 50-ft lengths are measured, the pins are transferred to the other ring. In that way, it is simple to keep an accurate count on the number of 50-ft lengths measured. This technique is helpful because one often needs to measure distances without assistance.

B.2. For appropriate elevation measurements (or differences in elevation), one can use a Locke hand level and either a 6-ft folding rule, level rod, or facsimile. If there is no one to assist, the opened 6-ft rule can be jammed or otherwise buried in the soil sufficiently to stand alone, and leveling along a line between two points can be done in the usual process of alternate backsights and foresights, using the Locke level as a substitute for the more usual engineer's level. For those unaware, the Locke level is a hand-held tubular device, about 8 in. long, which enables one to sight along a horizontal line to a level rod at the backsight and foresight locations. The procedure using the folding rule is a bit awkward, and it would be preferable to use a more suitable level rod and the assistance of a second person as rodman. However, there may be instances where one does not have the luxury of an assistant. In fill placement work, where I have used the technique, it is probably only necessary to determine elevations periodically to within 0.5 ft to keep track of the progress of the work for purposes of interim reports, and the Locke method, done with reasonable care, will serve that purpose. Final grades, of course, would be determined by more accurate, standard surveying methods, and most likely would be done by others (i.e., professional surveyors).

B.3. A small hand compass can be used to establish approximate directions. It also may be useful for locating shallow (buried) metal objects such as surveying markers or utilities, as long as one knows in advance their approximate locations by ties. A Brunton pocket transit can also be very useful.

For those not familiar with basic surveying techniques and tools, see any elementary surveying text. One of the very best (still), in my opinion, is *Elementary Surveying*, Vol. I, by Breed and Hosmer (1945). There are at least eight editions, with copyright dates starting in 1906! There are, of course, many new and extremely sophisticated surveying instruments and techniques now available, like laser beam technology, but for the kind of stuff I'm describing here, refer to Breed and Hosmer.

B.4. A few additional words about record keeping in general is warranted. As mentioned under "Clothing," one should always carry certain equipment on one's person at all times while engaged in field work of any sort. Note under B.4.b that I suggest and emphasize *no eraser.* This is because of the possibility that construction field notes or data may wind up as evidence in depositions or in court cases. Clearly, then, any erasures are suspect. It is good practice to avoid their use. When you inadvertently write down an incorrect word or number, simply and neatly cross out the error and add the correction. After a while, this procedure becomes a habit, and the value of your records is enhanced.

A second suggestion is to avoid ambiguous entries. On fill jobs, for example, use the letters "C" and "F" for cuts and fills, rather than "−" and "+" (requiring an interpretation). For approximate directions use "easterly" and "westerly," rather than "left" and "right." (On a highway location study, I once worked two weeks in the office plotting surveyor's topographic notes and preparing contour maps. His notes read "left" and "right" of centerline. He was out of the country on vacation, so we could not contact him for clarification of the ambiguous notes. Upon consultation with the project engineer, we agreed to guess which was meant. We guessed wrong, and two weeks of work produced upside down topography!)

One way of avoiding ambiguity is to adopt the habit of explaining symbols used with a note explaining the symbol's physical significance; for example, "+ means the volume is increasing." Another helpful technique is to assume that another person will have to interpret your notes or data without any opportunity to question you. If you consciously take this approach on a regular basis, after a while it will become habitual, and the result will be consistently clear, unambiguous notes or data. I have observed over many years of teaching laboratory courses, including surveying, that students develop bad habits in record keeping, because in their academic setting they will be interpreting their own notes or data, usually within hours or days, and so they subconsciously rely on their memory for interpretation.

Speed message forms are those commercially available snap-off forms with built-in carbons. They are very useful and time saving when you wish to mail short but important notes from the field and wish to retain a copy. The plastic envelope suggested for the pocket notebook is to protect against obliteration of important notes by perspiration or rain. This is also sometimes called a "nerd pack."

B.5. Many times in the performance of an assortment of field activities, especially inspection work, it will be necessary to measure horizontality or verticality, sometimes rather precisely. A 6-in. rule and a good-size, quality carpenter's level (one about 4 ft long is suggested) will enable one to measure quickly for horizontality. Incidentally, the measurement will be quicker, easier, and much less susceptible to error if the zero end of the rule is *physically* the end of the rule. On many pocket rules, this is not the case. A plumb bob and 6-in. rule can also be used to check plumb. The plumb bob has the advantage that it can be readily carried on the person.

C and D. The penetrometer, Torvane, and geostick are described briefly in Section 8.1.1.

Sample bags of various sizes and certain digging tools will be needed for a variety of purposes. The small, heavy plastic bags might be used for holding small samples scraped from the wall of a test pit for textural classifications. The large burlap bag and larger shovel would be needed to obtain a borrow sample for laboratory compaction testing. A garden spade would be suitable for digging field density holes or for obtaining small bag samples. The geologist's pick would be useful in many ways, with one specific use being the retrieval of jacked samples from the wall of a test pit (see Section 8.1.2).

E.1. Test pit sampling is described briefly in Section 8.1.2 and illustrated in Figures 8.4 and 8.5. The jacking kit listed is simply a wooden box containing the jack, wood blocks, base, a plastic plate, and tee. The tee is an aluminum, T-shaped part, the vertical section of which transfers the load from the jack piston to the hollow section of shelby tube. The horizontal portions support the tube (front) and the jack (back). The base is simply a wooden trough extending approximately across the width of the test pit, upon which everything rests. The wood blocks, of assorted arbitrary sizes are used as chocks to transmit the load from the rear wall of the test pit to the shelby tube section. The tube is thus jacked into the wall, periodically resetting the jack piston, changing blocks as needed, until the tube is fully jacked into the wall. The filled tube is then extracted by digging around its exterior with the geologist's pick. The entire kit is fairly heavy, but portable.

E.2. Field-density testing has, of course, been described extensively in prior sections. Detailed itemization of equipment has been omitted, however, because several options are available for the various operations. Note that under sieves, I list a full-size #4, and a full set of miniatures. The former is normally needed to determine the percentage of soil from the density hole that is retained on the #4 sieve, this quantity being needed to allow an empirical correction to the target value density. Recall that, because of the mold size, the #4 material is excluded from the laboratory test. If some larger size is excluded, utilizing a larger mold, then that sieve should be brought to the field. For example, ASTM permits ¾-in. particle sizes if a 6-in. diameter mold is used.

The full set of miniature sieves (about 3 in. in diameter) would enable one to do a sieve test in the field by hand shaking. Such testing is not normally done as a matter of routine, but it can't hurt to have that capability in the field, especially if the sieves are available and not needed elsewhere. Also, there are some circumstances when field sieving might have an important purpose. One would be an attempt to test the idea of the compaction data book that I proposed and described in Chapter 4, as a means of dealing with the problem of changing borrow. Plots of grain size distributions are needed to apply the method, and hand sieving would provide more accurate plots than reliance upon visual estimates of gradation. Another circumstance might be one where the field person is inexperienced and has not yet developed the skill of visual soil classification. In such a case, hand sieving could be used as a check (or substitution) for visual classifications. (For those unaware, with practice it is possible to prepare fairly accurate grain size distribution curves by visual and tactile inspection, utilizing techniques developed by Donald Burmister of Columbia University. See Chapter 9.)

It should be recognized that hand sieving in the field will introduce some error, inasmuch as it will not likely be possible to dry out and wash the sample in the sieving process, as is done in the laboratory. Without washing, fines will adhere to the coarser particles, thus *always* yielding erroneously low percentages of fines. If the quantity of fines is very large, the error may be significant. (It occurs to me that this problem would make still another simple applied research investigation to be added to those described in Chapter 5. The goal would be to develop a prediction capability for adhered fines corrections, to improve the reliability of the process used for the suggested compaction data book.)

E.3. Load testing in the field is often a very important, time-consuming, expensive activity. A single pile load test, for example, can take a day to set up, several days to run, and cost several thousand dollars. Plate load tests can be relatively simple, 5-minute tests such as that described by Giddings (Section 7.1.1), or larger tests comparable in elaborateness and cost to pile tests. It is beyond the scope and focus of this book to include details of large-scale load testing, but I offer some brief commentary about the role of the field person.

First, pile tests and plate bearing tests may both be thought of as plate tests because the load applied to the pile is transferred to the butt end of the driven pile by a thick steel plate, usually about 3 in. thick. Second, both tests measure loads and corresponding deformations. Loads are usually applied by a hydraulic jack, and a common method involves jacking against a dead load composed of tons of steel ingots piled on a reaction beam. The plate is instrumented in several ways to allow independent means of measuring the movement of the plate, three of which are (1) extensometers (Ames dials, typically) mounted at corners of the plate, (2) engineer's level readings (foresight rods attached to the plate, backsight rods on selected immovable benchmarks on site), and (3) piano wire strung in front of a steel rule mounted on the side of the plate.

It is commonly the function of the field inspector to approve the setup, and participate in the running of the test, principally by reading the instruments that measure plate movement and, of course, recording all results. The setup is done by the construction workers. This can take several hours, using cranes and the like. Details will not be included here, except to say that when they are done, you'll be asked, "Is it O.K.?" This can be rather sticky, for if you simply say "Yes," there's an inference that anything that goes wrong later may be blamed on you, with a "He said it was O.K.!" chorus from the contractor's people. To make matters more complicated, there is a good likelihood that the entire setup was done without your being present to observe the process. I recommend that, if at all possible, you make every effort to observe the process, at least intermittently, for it is much easier to spot and correct problems during the setup than after completion. For example, one item that should be checked, using the carpenter's level, is that the reaction beam is horizontal in at least two directions. A typical limitation on the allowable slope is ½%. (A code will likely govern this.) Clearly, it is much easier to correct an unacceptable slope early in the setup (by having shims inserted) than when the full dead load has been added. (I participated in one pile test where the dead load exceeded 350 tons.)

Some of the equipment that is used in the test itself, or that can be helpful in taking the necessary measurements, is listed under E.3. The items all pertain to the measurement of the movement of the plate under the influence of the load increments and decrements. Generally, increments (decrements) are added (removed) when plate movement has ceased under the influence of any given load. The test is run continuously, so inspectors work in shifts. A failed test is commonly defined by some limiting net movement of the plate (i.e., not including elastic deformation). In New York City, for example, a net movement of a steel pile exceeding ¾ in. defines a failed pile.

The instrumentation apparatus is commonly supplied by the contractor and is part of the setup. This might include the Ames dials, rods, piano wire, and the means for setting them up in a firm manner, that is, in such a way that they will not be physically disturbed during the test. (These are some of the other things that should be checked before saying that meaningful "Yes.")

E.3.a. In my experience, the agency that supplies the engineer's level and is responsible for its adjustment (and perhaps certifying that adjustment) is often left up in the air. I don't think it should be, and I recommend that this be settled in advance. What happens on a typical load test is that none of the independent sets of data agree, and difficulties (and even arguments) develop over which is more correct. With two foresight rods, a set of Ames dials, and a piano wire, there will be *four* sets of independent data. If the level is in proper adjustment, and is so certified in advance, the sources of discrepancy will be reduced. (Notice I say reduced; they're never eliminated.) Breed and Hosmer describe simple level adjustments, but often it will be necessary to send the instrument to experts for shop adjustments, especially since instruments used around construction projects are likely to have been knocked around. An alternative, of course, is to get another instrument. It seems to me foolish to risk messing up an expensive test by using a level badly out of adjustment.

E.3.b. Ames dials are usually mounted on three or four symmetrically located points above the plate, with their plungers initially pushed in. An arbitrarily set initial reading for a dial with a 1-in. range might be 0.026. As the test progresses, at some time late in the test (higher load), the plate will have moved to near the range of the dial, at which time a reset of the dials must be done—before the plate leaves the dials. Neat aluminum blocks are recommended for this purpose. One would simply write "reset dial" on the data sheet, read and record the dial (say 0.963), push the plunger upward, insert the aluminum spacer cube, and read and record the equivalent new dial reading (say 0.483). Alongside the "reset dial" notation, write 0.963 ≡ 0.483.

An alternative to this procedure would be to remove the entire dial and remount it at the new setting, but this is more cumbersome in that you'll need pliers or other tools, and there is greater risk of otherwise messing up the instrumentation. Remember, it may be dark, cold, and raining or snowing, and you will be typically working in tight quarters and in close proximity to all the other instrumentation. (If you hear a "spronggg," you just blew the piano wire system. Of course, only you and the Lord will know!)

E.3.c. A flashlight with fresh batteries will be needed if you are assigned a shift encompassing darkness.

E.3.d. Admittedly, it may be superfluous, but a hand telescope or binoculars may be helpful to give you remote capability to read the dials. In that event, a rather powerful flashlight will be needed for night work. One caution, however: The pointers on some dials have a tendency to get stuck on the inner face of the glass, requiring a judicious tap or two to unstick the pointer to get the correct reading; this is apparently caused by the accumulation of static electricity. You should check whether this occurs with the dials you are using. If it does, remote reading is not advisable.

The idea of remote readings is not a matter of comfort and covenience, but rather one of a further precaution against disturbing the instrumentation. As noted, the test area is usually pretty tight and cluttered, and heavy rain can create a quagmire or even standing pools of water to further complicate things. Incidentally, this is one of the other things that can be checked if you have the opportunity to observe the "dock wallopers" setting up the test: the room for movement provided in the immediate area of the jack and instrumentation and the grading (topography) of the area. By tactful requests or suggestions you may be able to make your job easier later. I would not recommend that you demand or in other ways appear to be supervising, for you may inadvertently create a potential for later culpability should something go wrong, and, as Murphy says, it will. Also, the "dock wallopers" (the heavy equipment operators who set up the test) may resent it if you come on too strong. Put your requests in the form of "favors."

8.2.1 Summary Comments

The list presented at the beginning of this section is based only on my experience, so it is by no means intended to be complete. But I think that the approach I describe is a sound one. I urge organizations to add to the list or to prepare their own, and to provide their field personnel, in a systematic way similar to that described, with the equipment and other backup that will maximize their potential for effective field performance. I suggest that individuals, particularly the younger, inexperienced field person, keep careful notes in the field, and regularly review and summarize their notes at the completion of a field assignment, with an eye toward improving performance on a similar future assignment. With apologies to Alexander Pope, "To err is human, but to screw up the same thing repeatedly is unforgivable."

8.3 CASE HISTORIES

To cover the remaining items I consider important in fill control field techniques and inspection, I decided to intersperse most of the commentary within descriptions of case histories, principally the one designated as Case Study 1, Section 4.1.1. Rather than repeat myself, I suggest that you reread Section 4.1.1 to refresh your memory regarding some of the essential features of the job, notably those dealing with the focus of Chapter 4: major problems in fill control.

An approximate site plan and profile (prepared from memory) is shown in Figure 8.6. The job, while very small compared to some described in Chapter 7, was the largest I have been assigned to or involved with (about 150,000 yards). With the possible exception of a short job assignment in northern New Jersey, which I will describe briefly later, it was probably the most problem-plagued job with which I've ever been associated. I hasten to add that this was not through any fault of mine or the company for which I worked. Indeed, it was through our efforts that the job was straightened out and "completed." (The use of the quotation marks will be explained in due course.) Interestingly, it has been my perception that one learns a lot more from these types of jobs than from those that run relatively smoothly. I hope I am right, and that you will agree.

The job was complicated administratively at the outset by an unusual layering of interested parties, due primarily to its location in Appalachia, an economically depressed area. Because of the very high unemployment rate in the area, caused largely by the demise of the coal mining industry, incentives were developed to attract other industries to the area. One such incentive was site preparation work at no cost to the prospective industrial client. That was the case on this job. Thus, to start with, there were three levels of government involved: federal, state, and local. Add to that the architect, design engineers, earthwork contractor, soils engineers, and, late in the job, even geologists. The administrative complications were considerable. Even though I would be

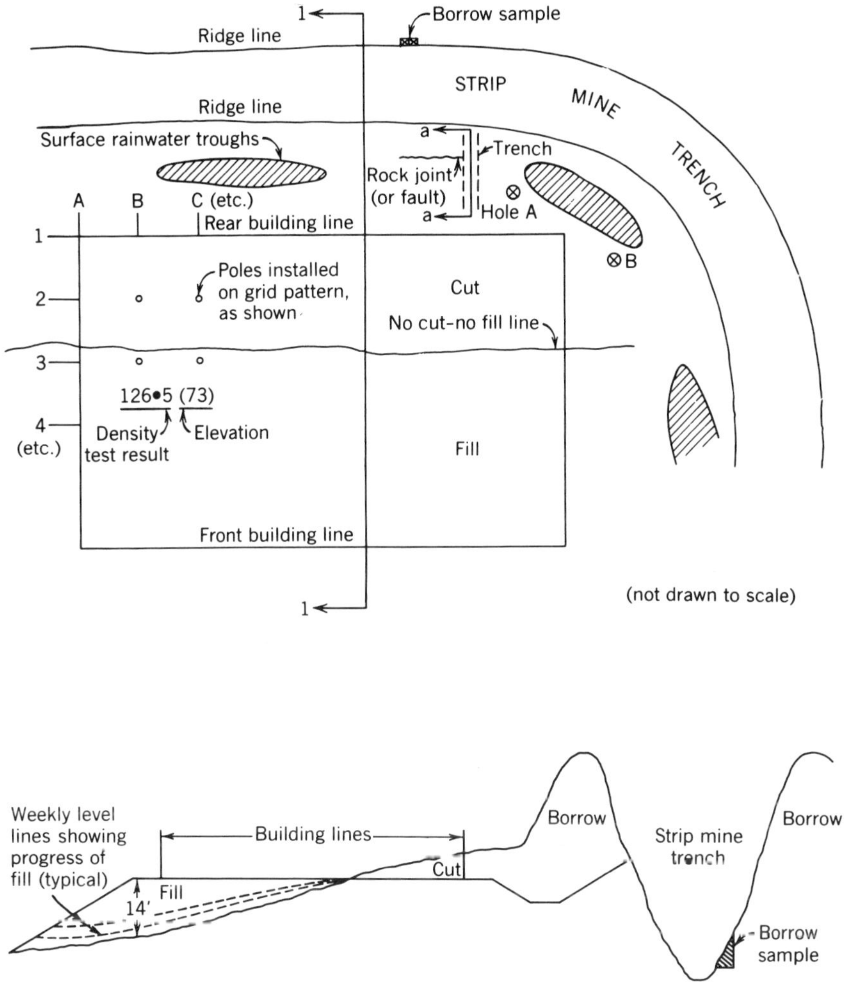

Figure 8.6. *Case Study 1.*

involved only with the fill inspection, and not with executive-level policy discussions, as the only representative of my company on site, I found myself making lists of the people involved in order to provide coherent reports to those to whom I was accountable. The list included affiliations and phone numbers. (This is a good early step in the record-keeping process on any job.)

As related earlier, I arrived in the late morning at the design engineer's office, and visited the site that afternoon. By evening I had become aware of quite a number of technical problems caused largely by the fact that the design engineers had called in our company much later than would have been ideal. (See Section 4.1.1 for summary.)

Before continuing with a description of the field problems on this case, and how they were resolved, you should be aware that I was in frequent telephone contact with the project engineer, at his insistence. While I was older than most inspectors would be, having been in the armed forces for 4 years, I was not much more than a geotechnical rookie, this being only my second summer's employment with a soils consulting firm. I did have earlier summer engineering jobs, but those were more broadly varied, including structural design and highway engineering. I had not yet obtained my PE license, being one year short of the experience required to take the test. I include this background information because it is typical for field inspectors to be relatively inexperienced. Thus, at least for small jobs, supervision is commonly provided by phone and, where possible, by short site visits by the project engineer. In many cases, then, the solutions to field problems are often augmented by the advice and instructions provided from the main office. However, there is a reasonable limit to how often one can or should call for advice. Deciding when to call, and with regard to which problems, is a problem in itself! In a nutshell, while some help is available, more often than not, you'll have to sink or swim by yourself.

Since filling had already started, one of the first problems I attempted to resolve was the obvious inadequacy of compaction capability, as reflected by the presence of a single compaction roller, shown in Figure 8.7. This photograph illustrates two problems written about earlier: inadequate compaction energy and a major imbalance between compaction and hauling–placement capabilities.

Figure 8.7. *Compactor, circa 1930.*

I first spoke with the earthwork contractor's superintendent about the two problems but shortly realized the best follow-up would be to prove the inadequacy. This was done by my first field-density test. On my suggestion, we both agreed to observe the placement and rolling process. The fill was brought in by the scraper (see Figure 8.8) and placed to a lift thickness agreed upon, about 12 in. This was measured and then rolled with (as I recall) five passes of the roller. Using the sand cone apparatus (Figure 8.3), I determined the dry density to be significantly less than the 95% *standard* Proctor called for in the specifications. (Note the emphasis. At this time, the major problem had not yet been discovered.)

The corollary problem of equipment imbalance, while not susceptible to a proof as explicit as the roller energy inadequacy, was fairly easy to resolve, since the superintendent was already convinced that he needed heavier compactors. Incidentally, one of the things I learned from this and similar experiences is the almost extraordinary ignorance of many earthwork contractors (or their employees) about field density testing.* As you will see by the extensive amount of heavy equipment that subsequently was brought to the job, this was a sizable contracting company. The conclusion I have drawn from the puzzled reactions of contractors to the appearance of field density apparatus

Figure 8.8. *A scraper or pan.*

*I do not use the term "ignorance" with any insulting or derogatory intent, making an important distinction between ignorance and stupidity. Thus, a very intelligent person can be ignorant of some fact or technique, as is the case and connotation here.

on the site, on more than one occasion, is that contractors have an "imbalance" themselves, in that they are heavily preoccupied with moving soil, but don't think too much about compacting it, perhaps an intangible major problem in itself.

After being convinced of the need, the contractor brought in a large number of Euclid trucks from another job that bad been completed nearby. With some discussion and consultation all around, an idea evolved to use loaded Euclid trucks with overinflated tires for compactors. As I recall, the contractor estimated that he could safely inflate the tires to about 70 psi. Figure 8.9 shows two such trucks in the process of compacting the soil. Although not a standard procedure, it worked well. One of the things that had to be monitored fairly often and carefully was the compaction of the "missed" portion between the rear wheels. The truck operators were instructed to move laterally the width of two tires (about 4 ft) after the required number of passes had been completed.

The trucks were also used to haul and place soil as shown in Figure 8.10. Power shovels (not shown) were used to load the soil in the trucks. Notice that the Euclid is in the process of dumping the fill, and that a dozer is required for spreading. Unlike the scraper, the trucks cannot be used effectively for placing the fill in a lift of uniform thickness. The dozer operation is an added cost. Also shown in the photograph is a sprinkler truck (background, behind rear of the Euclid). The time period of this job was rather dry and the borrow got unmanageably dry, requiring the added steps of sprinkling and mixing before compacting. Also shown in this view are scrapers.

Figure 8.9. *Euclid trucks as compactors.*

Figure 8.10. *General view of filling operations.*

After addressing the immediate problems of compaction and equipment imbalance, necessary to keep the job moving satisfactorily, the problem of locating adequate borrow of acceptable texture was next. As reported in Chapter 4, this problem had apparently not even been considered before our company was contacted. The soil for the filling that was being done prior to my arrival was obtained from the cut area toward the rear of the site (see Figure 8.6). As it developed, and was fairly evident from the beginning, the amount of fill required far exceeded the cut volume. First, the soil near the surface in the rear contained substantial amounts of organic material, including some rather large branches. Most of the latter were removed by hand and burned periodically*, and some additional stripping and wasting was necessary. Second, we hit rock at fairly shallow depths in the cut area.

Careful earlier planning would have included a search for suitable borrow, including studies of regional soil maps (what I call paper reconnaissance), physical reconnaissance, negotiations with the owner of a potential borrow, some physical investigation such as test pits or borings to determine the extent and texture of the borrow, and finally the actual steps toward acquisition. For several reasons, this sequence could not be followed. Most obviously, the job was underway and it was impractical to put everything and everyone on hold, especially because of the added factor of the unusual administrative complexities of the project. Also, as I understood it, available funding was tight or questionable. The upshot of it was that, after some rather quick local reconnaissance and phone calls, a decision was made to use the soil from a contiguous strip mine area for fill.

*It may be that burning or burying trees and brush will now be prohibited by environmental regulations.

Figure 8.6 shows the location of the strip mines that were used for borrow. Strip mining is a procedure where the ore (in this case, coal) is at such shallow depths that it is more economical to strip away the overlying soil and then mine the soil in an open pit operation. Deep veins are, of course, mined from vaults reached by shafts and tunnels.

Access to the trough of the strip mine was found during reconnaissance through a pass several hundred yards away. With the cooperation of the contractor's superintendent, I traveled to the area on a bulldozer and had the operator cut into the base of the strip mine to expose a section judged to be representative of the soil. Figure 8.11 shows the exposed wall. As may be judged from the photo, the composition of the soil was extremely variable, containing many obvious small boulders and cobbles, but also containing substantial quantities of all smaller sizes and categories: gravel, sand, silt, and clay. The high percentage of clay is responsible for the free-standing vertical face of the cut. Upon closer inspection and subsequent laboratory analysis, it turned out that the clay content was as much as 40% and was of low to medium plasticity. For most purposes, such a high clay content would be considered a poor characteristic for a structural fill. However, because of reasons already cited, the decision was made, with some reservations, to use the borrow. The advantage of proximity and easy availability, without hassle or acquisition problems (not to mention low cost), undoubtedly also contributed to the decision.

Figure 8.11. *Exposed wall of borrow area.*

The borrow was of glacial origin, a till deposited directly by the glacier. This explains the unsorted nature of the soil, resulting in the very wide range of sizes. The natural till, located beneath the excavated material from the strip mining operations, was very dense—a result of the large pressures exerted by the ice sheets that deposited the soil. The combination of cohesiveness and high density resulted in a soil that was difficult to excavate. A practical ramification was that scrapers had to be pushed by bulldozers during construction excavation.

A sample of the soil was cut from the face of the borrow area, as shown in Figure 8.11, and was brought to the laboratory for compaction testing and textural classification. The soil was cut from a vertical section of the freshly exposed face in order to obtain a sample that was as representative of the borrow as was reasonably possible to obtain. The sample weighed about 175 lb, probably somewhat more than necessary, but it's probably a good idea to be generous and get more rather than less.

Shortly after the problems of immediacy had been resolved, the real bombshell occurred. An engineer from the industrial tenant for whom the site was being prepared arrived at the site, having been assigned by his boss to visit the site to see how things were progressing. What transpired is described in Chapter 4, Section 4.1.1.

After the specification blunder was rectified by changing the target value to 95% *modified* Proctor, the job progressed fairly smoothly for a while. As things settled to relative normalcy, I endeavored to establish methods control criteria with the contractor by suggesting to him that we run a test strip "field experiment." By this time, we had the results of the modified Proctor test on the borrow, and so a target value density (95%) was known. However, I explained to the contractor that it would be advantageous to both of us if we could establish a lift thickness and number of passes (with the loaded Euclid trucks) that would give us the target value density. Being a reasonable man, he agreed, and we laid out a strip (on the site so it wouldn't be extra work), measured the lift thickness as 12 in. and both observed and counted three, five, and seven passes over three segments of the strip. I then did density tests in each of the three segments. (One of the tests is shown in Figure 8.3.) Upon completing the weighings and computations, I reviewed the results with the contractor. It turned out that five passes were not quite enough, and seven produced a density comfortably above the target value. Thus, we rather amicably agreed that six passes would be the typical method. I could now exercise some degree of control other than the density test. This did not become an official part of the specifications, but rather a reasonable agreed criterion. I emphasized to the contractor that the six passes would be valid only if no other conditions changed significantly: borrow texture, borrow moisture, compactor, and lift thickness. If it was judged that one or more factors did change significantly, new methods would have to be established, perhaps by another test strip experiment *and* maybe even another laboratory compaction test.

Changes in borrow texture, and suggestions for how to handle this problem were discussed in Section 4.3. Moisture problems and suggested solutions were described in Section 7.1.1, under weather restrictions. Compactors and the effects of compaction energy have been treated throughout. Suggested compactor types and lift thicknesses are listed in Table 6.1.

8.3.1 The Hole

About the time everything seemed to be running smoothly, an incident occurred that made me a firmer believer in both Murphy's law and Monahan's theory of the perversity of inanimate objects. A worker came running toward me exclaiming rather excitedly, "There's a hole, a hole . . .," pointing toward the rear of the site. I strolled to the area, and sure enough, there was a hole. I photographed the hole (Figure 8.12), thinking, "They'll never believe this at the office." The approximate location of the hole is shown on the site plan, Figure 8.6, hole A. (Another hole appeared a day or two later, also shown in Figure 8.6.) The 6-in. scale in the photo (always a helpful device for photo records) indicates that the hole was approximately 1½ ft in diameter. It was also rather deep, as evidenced by the fact that one could not distinguish the bottom visually, and a small boulder thrown in gave audible confirmation.

Because of the obvious concern that subterranean mining had also occurred at the site at some earlier time, or possibly that natural limestone cavities were the cause of the sinkhole, a decision was made to bring in a geologist to investigate. A brief description of the sequence of steps in the investigation, and the results, conclusions, and recommendations follows.

Figure 8.12. *The hole.*

Understand that I was not involved in any official capacity in this work. In fact, the filling operations continued, so I was otherwise occupied most of the time with my own duties. I must say that I did observe the proceedings with much interest, and, in fact, did a little surreptitious investigation and photographic record keeping on my own, mostly during lunch breaks, evenings, and weekends when no one was about. I did not confer at all with the geologist, after initially detecting an unmistakable disinterest to do so on his part, an unfortunate but not uncommon attitude of geologists toward engineers.

The first step was to bring in a dragline to excavate at hole A. In a few days a large hole was opened to a depth of about 25 ft, where a cavity was found with the unmistakable remnants of thick timber shoring. (Remember that no borings had been taken for this job, and I would surmise that old mining maps and records had been lost; either that or no one thought to search whatever records were available.)

During this time period, we got some rain and an interesting pattern of surface rainwater troughs formed, distinctly elongated and roughly paralleling the strip mine trench. This was definitive evidence of a pattern of surface subsidence, prompting (I surmise) the next logical step in the investigation sequence: a trench dug in a direction approximately perpendicular to the elongated subsidence troughs. A power shovel was brought in and such a trench was excavated, as shown schematically in Figure 8.6*a* and in Figure 8.13 (see the section a–a, 8.6*a*). As seen in the photo, a very distinct joint or fault

Figure 8.13. *The trench—a view of section a–a, Figure 8.6.*

was exposed. If you look closely at the top of the photo, you can see the third dimension of the joint on the surface. This surface crack was traced for 40 or 50 ft in a direction parallel to the troughs, the rear building line, and the strip mine trench and ridges.

With our fears mounting, the next step was to initiate a boring investigation to attempt to determine the extent of the problem. Figure 8.14 illustrates that this was done while the filling operation continued. The boring investigation (actually corings, as they are called in drilling in rock) was, of course, quite a logical next step. There had been almost no geotechnical investigation done before the filling operation had started.

Figure 8.15 shows one of the coring boxes, with the typical arrangement of the cored material from top to bottom. Note the shattered, broken nature of the rock nearer the surface caused by weathering. Most importantly, note the printed comment, "Void: 23.5 to 26.0."

As noted, I was not involved in any way with the investigation sequence and neither was the company for which I worked. Thus, I was not privy to all of the results, for example, of the coring operation. I heard informally that an initial decision and recommendation was to relocate the proposed structure a few hundred feet away, but I was told later that the structure was, in fact, built as originally planned. I did not get any final, definitive word on what actually was done, but I'm sure curious, even to this day. I hope some day to return to the site, perhaps incognito to look around.

The reasons I chose to include the descriptions of this section, fully recognizing that they have nothing to do with filling operations, is that I think the material is interesting and instructive to field workers, particularly the inexperienced ones, to whom this chapter is particularly directed. I think it illustrates the importance of field observations and the need for an ability and willingness to interpret and act on those observations. The best thing I have ever read on that subject, and which I strongly recommend, is "Art and Science in Subsurface Engineering," by Ralph Peck, *Geotechnique*, March 1962.

8.3.2 Keeping Records and Writing Reports

One of the most onerous but very necessary aspects of field inspection work is keeping notes and records, some of which will be needed for periodic reports, written and oral, to the office and to the client. It may be that a tiny fraction of these notes will turn out to be of special importance when something goes wrong. The trouble is you don't know which fraction in advance will prove to be significant. We could instruct an inspector to "write everything down" and be done with it. However, with the increasing tendency toward more and more lawsuits, such broad-brush, easy advice can lead to more serious problems than it may solve. So more careful consideration is necessary.

One way to view the problem is to separate it into three parts: how to keep fill records, where to take density tests, and what notes to write down or otherwise record (perhaps more importantly, what *not* to record). Let's take the easy parts first.

Figure 8.14. *Coring and filling operations.*

Figure 8.15. *A rock coring box.*

HOW TO KEEP FILL RECORDS The techniques that I adopted for the strip mine case study just described are, I believe, adaptable to many fill control jobs. First, obtain or prepare a large drawing of the site plan and profile (see Figure 8.6). Second, ask the contractor to install a Cartesian system of poles on a 100-ft grid, as shown. (If you look closely at Figure 8.9, and others, you will see some of these poles.) Mark each of the poles with its location (e.g., c4). When density tests are done, it will be relatively simple to locate oneself on the site by pacing from the nearest pole. The determination of the approximate elevation of the test can be done using the Locke level as described in Section 8.2. With a little care, I would judge that one can be accurate to within a couple of feet in plan, and perhaps 0.5 ft in elevation, which is, I think, accurate enough for the purpose. The coordinates, elevation, and density result should be recorded in your pocket notebook, and subsequently plotted on the plan, with the decimal point of the density value serving also as the plotted point. The elevation can be recorded in parentheses, as shown by example in Figure 8.6*a*. The technique described is helpful in that one can, at a glance, get a feeling for the general distribution of testing coverage, and adjust accordingly if some region of the fill is not being covered that should be.

A companion technique was the periodic running of a line of levels such as section 1–1 (Figure 8.6*a*) shown in profile in Figure 8.6*b*. On the job described, this was done (again using the Locke level) about once per week, usually on a Friday afternoon. The results were plotted, on successive weeks, on a profile, thus providing a simple graphic way of keeping the office and the client advised of the progress of the filling operation.

The job required weekly *and* daily written reports, which at the time I regarded as a little excessive, but inspectors don't make such decisions. I suspect that the unusual amount of reporting was caused by the equally unusual complications of the job. At any rate, copious and careful field notes were both essential and very helpful in preparing daily and weekly reports.

The circumstances of the job were such that the daily reports could not readily be typed. In such a case, the use of snap-off forms, with built-in carbons, can be very helpful and time saving. The retained copies are indispensable aids in preparing longer, formal reports at a later date.

As may be surmised from the photographs shown here, I strongly favor and recommend the use of an inexpensive camera for record keeping. Photographs provide not only a virtually objective and infallible record but may also save a lot of time and effort, especially when there is an unusual demand for written reports.

The use of a portable cassette recorder may also be a sensible option for record keeping and reporting. These electronic devices did not exist (at least at reasonable prices) when I did inspection work, but it occurs to me that their use might be feasible in some circumstances.

One last precaution: If the method of installing poles is used, it will be necessary to advise the contractor and the compaction operators to compact very close to the poles and to monitor this fairly closely. Without such insistence, the compactor operators are likely to miss the poles by a couple of feet on either side, resulting eventually in sizable areas of loose fill. If a footing happened to be built on such a spot, unacceptable settlement would likely result. One additional way of avoiding this event is to have the poles laid out with an origin such that no poles will be placed where footings are to be located. Unfortunately, footing location may not be known at the time of site preparation. Incidentally, the same logic holds if one is doing a test pit investigation of a site. Avoid digging test pits at a point where it is known a footing will be placed or is likely to be placed. For example, if the location of the corner of a proposed building is known, call for a test pit to be located at a point 10 ft outside the building, on an extension of one of the building lines.

On small jobs, the methods of record keeping need not be as elaborate as described, but on large jobs, a methodical approach is needed, supplemented by graphics and appropriate recording devices.

WHERE TO TAKE DENSITY TESTS The specifications may spell out the complete details of field density testing, including how many tests (per thousand yards, for example), and where to take the tests. Or the project engineer may instruct the inspector on how many and where. The other extreme would be almost no such instruction, in which case the inspector must depend solely on his or her own judgment.

If the project engineer has the responsibility and flexibility of deciding how to proceed, I think the decisions should depend largely on who will be doing the inspection work. If the inspector is judged to be experienced, intelligent, articulate, diligent, personable, but capable of forcefulness when necessary, I would stress more qualitative means of fill control, with only a minimum number of field-density tests. On the other hand, if the inspector possesses none of these qualities, have him dig holes all day.

My point is that one cannot do both at once. By qualitative fill control, I mean such activities as observing all aspects of the compaction process carefully, establishing methods, and conversing with and advising the contractor intelligently. If the inspector can do none of these well, then base the fill control on as many density test results as can reasonably be done in a day.

My preference, if I had the option, would be the former. Certainly, there would have to be some quantitative confirmation of the adequacy of the fill, in the form of some field-density test results, but I believe that the best overall job would result from taking advantage of the skills of the inspector by allowing him or her to spend a good deal of time on the job exercising those skills, I'd say maybe only 25% of the day digging holes.

Where to take the tests? There are many ways to decide. A very simple technique, which has the dual advantage of explicitness and lack of bias, is described in Appendix F of the Asphalt Institute Manual (MS-1) *Thickness Design*, entitled "Procedure for Selecting Sampling Locations by Random Sampling Technique." The technique involves six flips of a coin. An example is given by Truitt (1983, p. 225). The example relates to selecting a horizontal location, but the technique could easily be modified to selecting random elevations. The technique has the added advantage of being very quick.

Of course, there will arise the necessity of taking tests that *are* biased, and this is where the judgment and other qualities of the field inspector are brought to bear, particularly the one of forcefulness when necessary. When, for various reasons (some predictable and some unforeseeable), the inspector has good reason to suspect a fill is deficient, the courage to insist upon testing at that location is essential. This may require excavation, and the contractor is not going to like it. I describe these and similar problems in more detail in Section 8.3.3.

OBSERVATIONAL NOTES AND RECORDS The third category of record keeping is in many ways the most difficult. I have offered advice on some details of taking and recording load test data and field-density test data, and this is relatively easy because it deals with facts, usually of a numerical nature. But there will occur literally countless numbers of incidents, conversations, and discoveries in the course of a day that may not fit the description of "data" but that may be nonetheless important, and often much more important. I can only offer brief guidelines, some based on my own experiences, and some on the advice of others.

In a word, be as *factual* as possible.

Richards (1976) presents no less than 25 Rules of Conduct for the field engineer, and supports each with real, authentic case histories. I cite just two of them here, and add some commentary of my own. The first is "Cultivate Satisfactory Internal Relations" (Richards Rule 9):

> The field engineer should avoid any *written* criticism either direct or implied concerning members of his staff or organization. Detrimental comments by the field engineer concerning his own organization may seriously embarrass him when produced in court.

The emphasis on the word *written* is mine. Also, I think Richards' advice is good without restricting it to "internal relations" or "his own organization." Certainly, it is worse to put criticism of your own firm in writing than of another, for you may wind up being fired on top of being embarrassed, if the document becomes detrimental in court. But I don't see where written criticism of anyone should be done unless you make the conscious judgment that it can *serve some useful purpose.* For example, if you find a discrepancy, or a "chink," or even an outright blunder in a written specification, *phone* your project engineer first and raise the issue diplomatically. As an example of an external miscreant, if a contractor is regularly exceeding the specified fill thickness, *don't* write "every time I turn my back, he doubles the allowable lift thickness." Instead, write "three consecutive lift thicknesses were measured as 18 in., 20 in., and 22 in. The contractor was advised that this violated the specifications."

The latter constitutes a combination of my advice on being factual, and Richards' Rule 9.

As a hypothetical example of reporting a dilemma overlapping professional, ethical, and moral obligations, consider the following: You are a professional engineer of some experience. You pass a construction area (let's say in the New York metropolitan area) where you observe some men working in a 10-ft deep, unsupported trench in soil. What do you do?

This is a good one to kick around over coffee. I have done so with a colleague and we concur that one should (1) ascertain quickly who the controlling agency is, (2) call immediately and alert the agency to the safety violation (OSHA) you observed, and (3) follow up with a factual letter (*not* a sanctimonious diatribe or lecture).

The foregoing is hypothetical, but it could very easily happen. I shall never forget attending a Friday lecture in soil mechanics at Oklahoma State University on stability of vertical, unsupported soil excavations. Over that weekend, a worker was buried alive and suffocated on a sewer job on the edge of town.

And so one can see that the field inspector is constantly faced with a wide variety of decisions regarding what to report, whether to do so orally or in writing, and what words to use, particularly if the choice is a written report or notebook entry. It is not easy.

Richards' Rule 25 is

> Maintain Adequate Records: The primary task of the field engineer is to be able to reproduce at a later date project events. Adequate records of project construction activity, conversations, visitors, telephone messages and other related project events will be maintained by the reactive field engineer.

That's quite a mouthful. Like Richards, I can't define exhaustively or precisely what "adequate records" are, or what "other related project events" might be, but the above advice combined with the caveat to be as factual as possible, will hopefully be of some help.

Finally, when in doubt, talk to your project engineer or another more experienced person whose judgment you respect before putting something in writing. But if you stick to facts, you'll rarely be in doubt.

8.3.3 Contractor-Inspector Relationships

Another sticky problem one has to deal with is that of contractor–inspector relationships. In the previous section, I described the ideal field person, using the words personable and forceful (when necessary). The reason such characteristics are important is that the field person has to be able to deal with all kinds of contractors equally as well, in terms of ensuring that the specifications are followed. Conversely, the contractor has to deal with all kinds of inspectors and other problems so that, as Morris (1959) has written, he "can complete his work with pride and still make a reasonable profit."

Before proceeding further, so that no one will get the impression that I am unduly biased, let me say that it is my opinion—and I believe that it is the only one that makes common sense—that there are in this imperfect world of ours "good" and "bad" people in every profession and walk of life: engineers, lawyers, clients, contractors, and inspectors. And we must try to deal effectively with each other. Because of the fact that my experience, as it relates to writing this section, has been as an inspector dealing with contractors, it follows that most of what I have to say is about these relationships. And despite the fact I have not had extensive experience, I believe I've been lucky (if that's the word) to have had experiences with a wide variety of people: the good, the bad, and those in between. In addition, I take a lot of notes! But you be the judge.

The reason I chose "personable" and "forceful" as pertinent adjectives to describe the good inspector is that these traits are needed to deal effectively with the "good" and "bad" contractor, respectively. An example of a good contractor was the one I dealt with on the strip mine job that I described earlier. (Actually, he was the earthwork contractor's superintendent. Throughout this section, I shall use "contractor" to mean the person in charge at the site with whom I dealt directly as an inspector.) He was good in the sense of Morris's description in that he had pride and a natural wish to perform well and make a reasonable profit. This was fortunate for me, and we worked together very well, had a thoroughly amicable relationship, and got the job done effectively in spite of a most unusual set of problems.

One of the contractor's deficiencies, however, was a surprising ignorance of even the rudiments of soil compaction technology, notably field testing and how it relates to the whole fill control operation. I found this to be a rather common deficiency among contractors. So, unless my experience was somehow not representative, and unless things have changed drastically, an inspector should not be surprised by this finding.

This lays the groundwork (no pun intended) for my first suggestion for dealing effectively with such a contractor. *Make a conscious effort to find specific, legitimate ways of helping him perform well and increase his profit.* I can think of three ways this was accomplished in Case Study 1. First, when the test strip density testing was done, I showed the results to the contractor, and we agreed together to use the interpolated result of six passes. Another approach would have been for me to simply state "we'll need seven passes." While I never discussed this with the contractor, I truly believe that he thought "Gee, he's not out to do a job on me," or words to that effect. Subsequently, on one of the few rainy days on the job, I suggested that we could keep working effectively if we immediately compacted the soil upon placement, that even a 5-minute delay in the fairly heavy, steady rain would wet the loose lifts to such a degree as to make compaction impossible. He agreed, and we converted another Euclid truck to compaction duty and were thus able to work a couple of more hours. The third suggestion was to use rock excavated from the rear (cut) area of the site as fill, breaking the rock with passes of heavily loaded tandem sheepsfoot rollers, followed by a lift of soil borrow to fill the void spaces in the rock. The initial thought was to waste the rock fill.

Each of these suggestions was well received by the contractor, for obvious reasons, and each was regarded as compatible with the specifications and toward the effective and economic completion of the job. Taken individually, none of the suggestions was a big deal, but collectively they served the purpose of harmony and maybe even a little pride all around. With all of the unusual problems of this job, I hate to think of what it would have been like if the contractor had been a "bummer." I believe that if you perceive the contractor to have the characteristics described, the approach and attitude of consciously seeking ways of helping the contractor with sound advice will pay large dividends. He'll probably be unaware of this strategy unless he also reads this book from cover to cover, and I don't expect that every earthwork contractor and superintendent is going to rush out to buy this book—though that would be nice. I am not suggesting here that the inspector curry favor with the contractor, but rather that he foster harmony by offering legitimate, sound, money-saving advice.

And now for the other side of the coin. I was assigned briefly to a fill job in northern New Jersey. While I wasn't familiar with the details of its administration, I learned from conversations that it was a rather splintered job, involving a general contractor and an unusually large number of subcontractors, one of which did the earthwork. Coordination and scheduling on such jobs is difficult under the best circumstances, but on this job the circumstances were a nightmare. Early on, the job was plagued by criminal vandalism, including sand poured into gas tanks, and vertical number 9 reinforcing bars, encased in footings, smashed into U shapes. As a result of the early problems, the schedule of construction operations got completely out of control.

As described earlier, the general contractor requested an unusual procedure to attempt to improve or restore scheduling; he asked for approval to raise the perimeter of the main structure to bottom of footing grade, to allow construction of the exterior footings. Although this was not good practice, since it would create, in effect, a diked interior, whoever was in authority approved the request, I suspect in sympathy with the general contractor, a good man in a bad situation. The idea was that the interior would be filled later. (With the general contractor's luck, I half expected a major rainstorm before this could be done, making a king-size lagoon of the site.)

Because only a narrow perimeter strip would be filled, the elevation of the fill could be raised rather rapidly. To further speed things along, the earthwork contractor got approval to bring in a lot of extra equipment, mostly scrapers, on a Saturday. I shall never forget that Saturday.

Earlier in the week, when filling of the perimeter was being done at only a moderately accelerated pace, I was already having difficulties with the earthwork contractor, principally with excessive lift thicknesses. I spoke to the contractor about it repeatedly, and soon perceived that he was "trouble." You'd get a "Yeah, OK," but as soon as you would turn your back, he'd signal to the scraper operators to "lay it down." I often wished that I had had a long-lens camera to catch him with his hands separated from his knees to a point above his head, in the posture of signaling the scraper operators. I had called the office about the problems, having recognized that I was dealing with an incorrigible. In such cases, a call from the office or a visit to the site by the project engineer can have some positive effect. On Saturday, the situation got almost out of hand. As noted, extra scrapers were brought in, but no commensurate extra compacting equipment. Scrapers were whizzing by, and more than once I was engulfed in dust clouds, and was actually a little nervous about being run over. (Scrapers have been known to tip over. See Section 8.3.4, following.) As you might expect, the contractor was busy with lift thicknesses, and the day was filled with arguments and recriminations. This was to be my last day on that job, as I was reassigned to another.

In a situation comparable to the one just described, an inspector has to muster the courage and forcefulness to assert himself, But do it in stages. First try to handle the problem yourself, rather than calling the office at the first sign of trouble or an uncooperative contractor. Keep written, *factual* records of conversations, violations, and assertions of violations transmitted orally to the contractor. At some point, inform your project engineer and seek advice and, possibly, assistance. One of the approaches I have heard about is to threaten to walk off the job, and then do it. This would be done only as a last resort, and when other circumstances are compatible with the potential efficacy of such action. For example, if the specifications or some regulating agency require certification before payment, you can make the point that the fill cannot possibly be certified if you leave the site. Such pressure can work wonders.

Assertiveness and forcefulness may also be necessary in inspecting fills in confined areas such as trenches, and behind basement walls, where large thicknesses of fill can be placed (or dumped) in a short period of time. The specifics of these problems were discussed more fully in Section 6.5.2.

INTERPERSONAL RELATIONS It is important that the inspector not become overly friendly with the contractor on a personal level, to the point of socializing. I made this mistake once. On a drilled pier job, where it was my responsibility to inspect the rock surface at the bottom of the belled pier for proper size and rock quality, I got on friendly terms with the drilling contractor. He invited me to a cookout at his home on the weekend, and I accepted. This was a job where commuting was not feasible, and I stayed in a hotel in the area, so a weekend cookout was an attractive invitation, A couple of days later, I was compelled to reject a pier bottom after the drilling rig had already been moved to another location on the site. This rejection was prompted by the fact that the rock was of the unusual type that would deteriorate very rapidly when exposed, as I discovered, even overnight. What happened was that I had inspected and approved the pier bottom in question late one afternoon, but the concrete delivery service had quit for the day. Upon reinspection the next day, the rock was found to have deteriorated to the point where you could pull out chunks, with some effort, with your hands.

I remember two things from this incident: my reluctance to tell the contractor about the need to redrill and the look on his face when I did. (To maneuver the rig back onto the hole is often problematic and always time-consuming, so the contractors are not happy to do so.)

What I learned from this is not to get too friendly with a contractor, because even if you do so innocently and with no intent to have it affect job decisions, there will be an almost unavoidable influence when it becomes necessary to reject some work requiring corrective action. Similarly, the look on the contractor's face suggested to me that he thought "I just fed you a huge steak, and you do this to me?"

Richards describes some authentic case histories regarding matters such as accepting direct favors and gratuities ranging from bottles to ball games to business lunches. His Rule 2 is, "Constantly be discreet. Never accept favors or gratuities." He also describes cases of outright bribes, some of major magnitude and national notoriety.

The opposite extreme of the friendly inspector is one who is rigid, cold, and totally impersonal in dealings with the contractor. This type can be regarded by the contractor as a nitpicking pain. Such an extreme stance is, in my view, not a good idea either, for it can impede the development of harmony such as described on the strip mine job. And sometimes the cost of harmony is to give a little. One last personal experience will illustrate what I mean. I was assigned to a pile driving job, and the specifications called for a cushion block for the butt end of the pile to be a block of wood cut from a tree trunk and trimmed to fit. (This is to prevent damage by the direct impact of the hammer.) The contractor instead used split chunks of wood, as I recall, each about as big as your head; before the start of driving, a worker would pitch a number of these into the cylindrical space above the pile butt. For several days, I observed this procedure and raised no objection to the procedure, despite the fact that I was well aware that it was, technically, a violation of the specifications. In due course, one of my bosses saw or heard about this procedure, and became very upset because of the violation and my allowing them to use "chips" for the cushion block. Not being bashful, and perhaps a little imprudently, I argued with him privately that there was, in my view, no practical significance to this variation in procedure, that after two or three initial impacts of the hammer, there would be no practical, distinguishable difference between the cushion block material, chunks vis-à-vis a one-piece, trimmed block. I pointed out that I would not countenance the practice of throwing blocks into the rig near the end of the driving of the pile (called fetch-up), as this could seriously diminish the actual carrying capacity of the pile. (A contractor doing the latter would be guilty of intentional cheating, the purpose being to reduce possible damage to the pile hammer by improperly cushioning the pile during fetch-up.)

What we had here was a difference of opinion on two issues: the enforcement of specifications and whether or not the one-piece block or the use of chunks, as described, made any significant difference. I believe the latter could never be determined in any scientific way, and thus will have to remain a matter of opinion. However, the broader and more important issue of specification enforcement is worth considering. It is no doubt the safer and more conservative approach to adopt a "specs are specs" rigid approach, since no decisions would ever have to be made. I admit that it may be controversial, and in some circles almost heretical, but I believe that one always has to make adjustments: not, "Are the specs being followed?" but "What is best to produce the best job?" No specification ever written, or that will ever be written, is perfect. Indeed, as indicated by the case studies in Chapter 4 (especially Case Study 2), many are very badly written or pieced together by people with little knowledge of the technology, necessitating major changes, legally or illegally. If a minor variation is permitted, and the result is better harmony on the job, what's the harm? The alternative can be hostility, day in and day out, between the inspector and the contractor, and I can't help but believe that this will somehow, but with certainty, result in a poorer job.

In retrospect, as an inspector, what I should have done was call the office about the chunk issue when I first observed what the contractor was doing, offered my opinion, and then accepted their decision, irrespective of whether I agreed or not. But frankly, at the time, I didn't think it was an important enough issue to bother the office. I don't remember the name of the boss, and he doesn't know this, but I have referred to him as "Chips" lo these many years. Who says field work isn't fun?!

8.3.4 Odds and Ends

Two items of some importance that did not seem to fit neatly into the chronological description of the strip mine job are included to close this chapter. One deals with personal safety, and the other with rock excavation.

A NOTE ON PERSONAL SAFETY Figures 8.8 and 8.16 show scrapers placing soil and picking up soil, respectively. Notice in Figure 8.8 that the scraper has a high center of gravity, as evidenced by the concentration of mass in the area immediately above and behind the driver. Now examine Figure 8.16, and imagine the scraper coming down the slope at about 35 mph or better. The combination of three factors can contribute to a significant chance of rollover: the high center of gravity; steep, erratic slopes (perhaps with partially buried boulders); and the "cowboy" attitude of many scraper operators. The latter explains the caption of Figure 8.16.

I spoke with a scraper operator about this on one job and learned that, indeed, they are known to go over occasionally. He told me of one such recent rollover that he knew of where the operator sustained a serious head injury. Some other interesting tidbits: Some scrapers are made with cabs that serve as roll bars, preferred by the older, heavier, or otherwise less agile operators. The younger, agile types choose the cab-free type, the better to jump clear if the occasion demands. I was also told that there are no *old* scraper operators,

Figure 8.16. *A scraper rodeo.*

because the kidneys and other inner organs can't take the abuse after about age 35, presumably assuming 15 or so years of prior experience operating a scraper. In Figure 8.16, note also the bulldozer assisting the scraper with the excavation in the dense soil.

A NOTE ON ROCK EXCAVATION As the cut proceeded at the rear of the site of the strip mine job, rock was encountered at shallow depths beneath the glacial till, necessitating both ripping and blasting. The shallower rock yielded to the ripping operation shown in Figure 8.17. This was the rock that was placed in the fill with alternating layers of soil fill, as described earlier. As we got deeper in the rock, the rock quality increased (see Figure 8.15), and blasting became necessary. Fortunately for purposes of cost containment, only one blast was required. See Figure 8.18. (Since I took this photo from the next county, I think you will agree that my timing is to be admired more than my courage.) Apparently this loosened the rock sufficiently to resume ripping effectively. These photos and the accompanying comments are included to further emphasize and implant the importance of rock excavation as a construction cost.

Figure 8.17. *Rock excavation by ripping.*

Figure 8.18. *Rock blasting.*

8.3.5 Quiz Answer

Yes. There would be no differential settlements only if all footings were of the same size, which would require that all column loads were the same, a virtually impossible case. See Section 8.1.1 for further description.

8.4 GLOSSARY

Tare The container in which a material is weighed; hence tare weight.

Tectonic (forces) Internal geologic processes that influence surface events. See any modern book on physical geology for an update on the new and emerging discoveries in this fascinating field.

Varved clays A series of repeating thin soil strata, each series representing a one-year cycle of alluvial (lake or *lacustrine*) deposition. Clay laminae (thin layers) represent winter deposition, and subsequent sand–silt layers represent the following spring (runoff) deposition.

CHAPTER 9

Septic Systems

Some of the most important fills are those associated with septic system construction. These "on-site wastewater disposal systems" are common in rural areas where population densities preclude the construction of "city sewers." Their importance can be most simply emphasized by the fact that about one-third of all homes in the United States use this method of sewage disposal. One in four new houses being constructed incorporates some such system. Factors that suggest the future importance and increasing use of such systems are (1) increasing concerns about environmental matters, notably the pollution of groundwater, (2) decreasing availability of funding for sewer projects, and the need to repair existing sewers, and (3) the surprising fact that one-quarter of the land area of the United States is populated by fewer than two people per square mile.

Despite these statistics, there does not seem to be a corresponding amount of attention paid to research and development of new and improved technology relating to septic systems, especially for residential use. A good example of this is afforded by examination of a recent edition of one of the most respected books dealing with seepage analysis: *Seepage, Drainage and Flow Nets*, by Harry Cedergren (1989). I had used this book for many years in teaching graduate courses in seepage analysis. When the third edition was published, I noted that a new chapter had been added dealing mostly with waste disposal. Alas, I was to discover the coverage was for toxic wastes, nuclear waste, industrial waste, and mining wastes . . . but not a word about plain old-fashioned toilet flushing! I surmise and suspect that most people do not consider the pollution potential of the individual residence to be of much import compared to the four sources covered by Cedergren, but I hope that this book will make the case effectively that the *collective* contribution of residential septic systems is sufficient to warrant comparable attention.

In addition to the pollution potential, I hope to also show that there is a corollary cost of neglect measured in many ways, largely associated with the broad term we might call "effective and responsible land usage" for such major national endeavors as construction and farming. In the scheme of things, these activities are, of course, of enormous importance. In addition to monetary costs, there are many intangible costs involved, notably the aggravation and frustration that is associated with the present regulatory processes.

Lastly, I would like to suggest to some young geoenvironmental engineer who hopefully will read this chapter, that he or she seriously consider becoming expert in both seepage analysis and those subjects that might be lumped under a proposed new umbrella term: Pollution Science. It is my observation that there are many people expert in seepage (e.g., Cedergren), and some people expert in "pollution sciences" (some will be cited subsequently), but not one (that I have discovered) who is expert in both, at least one who has so far been inclined to share that expertise in published works. Attention, young geoenvironmentalists out there!

Compounding the problem is the virtual absence of any means of disseminating useful information to the public regarding the maintenance of septic systems, particularly that of a preventative nature. Thus, many residential systems fail because of neglect and abuse, rather than because of any design flaw or construction deficiency.

Before the homeowner moves in, however, there is a series of sequential steps in the process, all of which are heavily regulated by government at several levels. Unfortunately, there are many deficiencies in the regulatory process, both in the written regulations, and in the manner in which the regulations are administered. This chapter is a review of much of the existing technology, and includes commentary about some shortcomings of the technology and regulations pertaining thereto. Suggestions for improvements to both the technology and the regulatory process are offered.

9.1 BACKGROUND TECHNOLOGY

Figure 9.1 shows three types of septic systems. These were taken from a publication of the United States Environmental Protection Agency (EPA), entitled "Small Wastewater Systems: Alternate Systems for Small Communities and Rural Areas" (1980), an excellent graphic treatment of about 20 alternative systems on one foldout sheet that provides a brief but comprehensive overview of such systems. Figures 9.1*a* and *b* represent what are commonly called standard septic systems, gravity flow systems wherein effluent from the house is discharged into the septic tank where solids accumulate by settling to form a sludge, Figure 9.1*a*. (This first step in the treatment process is analogous to what is called "primary treatment" in sewage treatment plants in sewered areas.

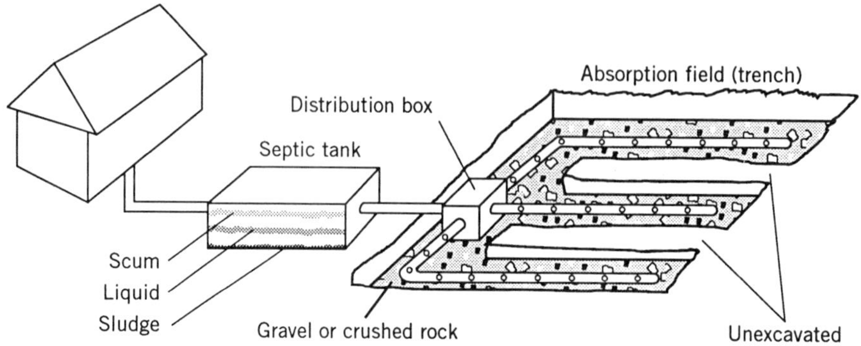

(*a*) Trench

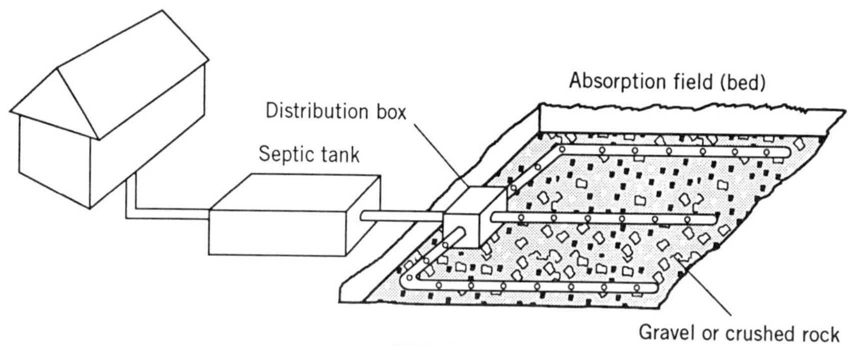

(*b*) Bed

Figure 9.1. *(a) Septic tank and soil absorption field—*trench. *Sewage bacteria break up some solids in tank. Heavy solids sink to bottom as sludge. Grease and light particles float to the top as scum. Liquid flows from tank through closed pipe and distribution box to perforated pipes in trenches, then flows through surrounding crushed rocks or gravel and soil to groundwater (underground water). Bacteria and oxygen in soil help purify the liquid. Tank sludge and scum are pumped out periodically.* Most common onsite system *(sic) Level ground or moderate slope. (b) Septic tank and soil absorption field—*bed. *Similar to (a) but smaller field. Total field excavated. Used where space is limited. Nearly level ground. (c) Mound system (used with septic or aerobic tank). Liquid is pumped from pumping chamber to perforated plastic pipe in sand mound that covers plowed ground. Liquid flows through rocks or gravle, sand, and natural soil. Mound vegetation helps evaporate liquid. Rocky or tight soil or high water table.*

Floating on the liquid surface in the tank is an accumulation that is commonly called scum, but which I prefer to call "slick", inasmuch as this term connotes the extensive presence of oil . . . as in "oil slick." The adoption of this term has, I think a useful purpose with respect to septic system maintenance, as will be seen. When the fluid level rises to the outlet port, the fluid

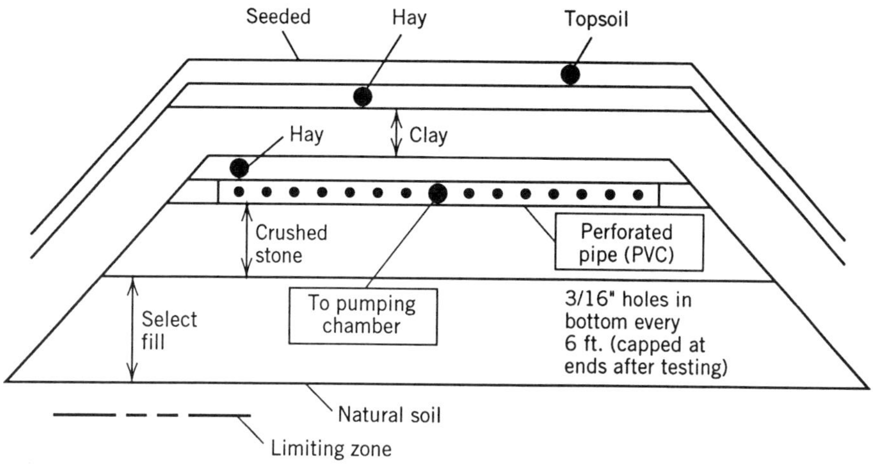

(*c*) Mound System

Figure 9.1. *Continued*

flows by gravity through a pipe to a distribution box, and thence to the absorption field through perforated pipes. (The absorption field is also commonly called a drainage or leach field.) The fluid then percolates downward through crushed rock and thence into and through the natural soil below.

It is in this last medium that the cleansing action takes place by virtue of a filtering action, removing bacteria and some viruses. Thus, it is necessary that a sufficient thickness of suitable natural soil exists below the field for that to occur. It is somewhat of a simplification, but nonetheless instructive to note that the natural soil should be silty. Percolation through such soils will result in the most effective filtering action, and will thus remove enough bacteria and viruses to render the resultant effluent "safe." If the soils are predominantly clayey, the fluid cannot flow rapidly enough to accommodate the discharge from the house, and if the soil is too porous (gravels and coarser sands), the flow will be too rapid for cleansing to take place.

For standard septic to be acceptable in most jurisdictions, the depth of suitable soil beneath the drain field and the *limiting zone* must be 60 in. This requirement is usually determined first by a visual inspection of test pits (also called "probes"), followed by a series of percolation (or "perk") tests. It is important to note that the test pit inspection must pass before a percolation test can even commence. The limiting zone is defined by regulators as either the water table or some form of aquaclude: a barrier through which water cannot pass, such as bedrock or some soil formation that is virtually impervious (clays, or "fragipans").

If the depth of suitable soil is between 60 and 20 in., a mound system is generally acceptable (Figure 9.1*c*). As the name and figure indicate, suitable soil is brought in and placed to form a mound (often called a "turkey mound"), on top of which is built the drain field. As shown in the figure, it is necessary to provide a storage tank containing a pump so as to feed effluent from the tank to the higher drain field. If less than 20 in. of suitable soil is found during the test pit inspection, or if the perk test fails, neither a standard nor a mound system is acceptable, and some form of experimental system or approach will be required. In such a case, the regulatory body (in Pennsylvania, the Department of Environmental Resources, DER) would have to approve the design on a case-by-case basis.* Typically, it would be necessary and desirable to retain a specialist in septic system design, generally a registered professional engineer or sanitarian whose specialty is sanitary or environmental engineering. The experimental design would likely be one that incorporates aerobic treatment in the septic tank. (Standard systems involve anaerobic treatment, that is, treatment in the absence of oxygen.)

Other forms of "high-tech" systems exist or are being developed. As might be expected, the cost of systems such as these can be substantial, indeed, perhaps prohibitively so. In such cases, the best decision may very well be to consider the lot "unbuildable."

Recently, the protection of wetlands has become a major factor in siting construction projects of all sorts. Indeed, all forms of land usage, including farming, have come under increasing scrutiny by environmental groups and regulatory agencies. Further commentary about this new and controversial matter will be included in subsequent sections.

Before proceeding with a more detailed consideration of the sequential steps of septic system technology, a few general observations are appropriate. Although perhaps trite, the adage, "A chain is only as strong as its weakest link," applies very well to this process. And there are many weaknesses in current practice. I shall attempt to point these out as descriptions are presented. Some are technological, some have do with the written regulations, and still others have to do with the manner in which regulations are administered, and by whom. In many cases, the weaknesses are a result of a combination of all of the foregoing.

The four steps in the process are (1) site investigation and evaluation, (2) system selection and design, (3) construction, and (4) maintenance. I believe that the overwhelming number of weaknesses are in the first and last steps: site

*Throughout this chapter, references will be made to practices followed in Pennsylvania. To what extent these practices will be similar to those of other states is not known, but certainly there should be some similarity, inasmuch as there is a single federal agency that dictates, to some degree, what *minimum* standards must be followed so as to be in compliance with federal statutes. This agency is, of course, the Environmental Protection Agency (EPA). To attempt to compile comprehensive information about all of the states' practices would not be sensible, for two reasons. First, the document would likely be of book length and take a great deal of time and effort to prepare, and second, because of turnovers in administrations, it is inevitable that many of the practices and regulations would have changed before the book went to press! As will be seen, almost all aspects of septic system technology, and the inherent controversies that evolve, are changing constantly.

evaluation and maintenance. As an example, I attended a seminar at a development in the Poconos in 1990 that was offered by a professor from a major university and attended by a large number of homeowners. While what he presented was well prepared, correct, and informative, he completely omitted step 1, and barely touched on step 4, that of maintenance. Since the audience was composed mostly of homeowners, this was a most surprising and glaring omission. The attendance confirmed what I had long believed: that people who own septic systems are hungry for information about them. Conversations that I had with some participants (and many others before and since) were revealing to say the least. One person had not had the septic tank pumped out in 20 years, and another did not even know where the tank was! So much for maintenance!

9.2 SITE INVESTIGATION AND EVALUATION

In Pennsylvania, the first step in the investigation process (after of course the necessary permit and fee), is to retain someone to excavate test pits at potential locations on the site where the eventual drain field might be located. The local Sewage Enforcement Officer (SEO) is apprised, and does a visual inspection of the test pit (probe), with the principal focus of attention on "depth to limiting zone," an expression analogous to depth of suitable soil, as described in the previous section. More often than not, the limiting zone is defined by the level of seasonal high water table (rather than by some aquaclude such as bedrock, an impervious layer, or fragipan). If the water table is too close to the bottom of the intended drain field, there is an obvious potential for pollution. The typical criterion for determining this limiting zone is the presence of *mottling*. Indeed, this condition is almost always used as the basis for rejecting a site for standard or mound systems. As noted earlier, 60 in. is needed for standard, and 20 in. for mound approval.

Mottling is a phenomenon that develops in certain soil types that is caused by cyclic (seasonal) fluctuations of the water table, the mottling (if it develops) being visible in the zone between high water table and low water table during the annual cycle of the rise and fall of the water table. The mottled appearance is caused by the chemical reactions of oxidation and reduction that take place in certain clay minerals during low and high water table, respectively. The resulting appearance is a certain blotchiness: a series of somewhat circular reddish and gray blotches.

Herein lies the first in a series of what I have referred to as administrative weaknesses, in this case with both the written regulations (in Pennsylvania), and the manner in which the regulations are currently enforced. The section of the written regulations that creates major problems is here quoted exactly as it appears in the rules and regulations promulgated by the "Sewage Facilities Act," P.L. 1535, 24 January 1966, Title 25, Pennsylvania Code, Chapter 73, Standards for Sewage Disposal Facilities, Section 73.1: "Limiting Zone—Any horizon or condition in the soil profile or underlying strata which includes . . . (i) a seasonal high water table, whether perched or regional, determined by direct observation or indicated by mottling. . . ." The aspect of this regulation that creates major problems is the absence of some qualifying language referring to the presence of mottling. I believe that some descriptive adjectives are badly needed, such as "continuous horizontal mottling," or "pervasive mottling delineating the top of the limiting zone." The way the present regulation is written, the presence of a single mottle, or what some regulator *says* is a single mottle, above the 20-in. depth required for a mound system can, and very frequently does, disqualify the site for standard or mound septic systems. In such cases, that ends the process there and then, and the lot is declared unsuitable (except, as earlier noted, for some expensive experimental system).

The principal reason for the problem associated with the present language of the regulation is attested to by the fact that, without the recommended qualifying adjectives, it is virtually impossible to get agreement, *even among experts*, as to the presence and location of mottling. The case histories that follow will, I believe, illustrate the point.

Case History 1 A probe inspection was done on a lot by a soil scientist who has a Ph.D. from the University of Wisconsin with a major in soils and a minor in geology; has extensive research and teaching experience (Penn State); has 82 publications in scientific journals; and has a consulting business subtitled Professional Soil Scientists. His assessment was that the limiting zone, as defined by the presence of mottling, was at a depth of "about 20 inches on one side of the pit, and at about 17 inches or so on the other side." The local regulator, with the support and assistance of state personnel, ruled that the test pit "had mottles at 10 inches." On the basis of the latter assessment the Permit Application for a Sewage Disposal System (for standard or mound) was denied, thus rendering the five-acre lot "unbuildable."

Case History 2 In 1983, a developer in the Poconos brought in a consulting geologist from Lehigh University to examine a series of lots in a section he was developing. The geologist's report stated: "All sixteen lots examined contain soils suitable for on-site sewage disposal" (Ryan, 1976). Since that time, based on the mottling criterion of the regulation cited above, the section has been consistently rejected by regulators. In one instance, a magnifying glass was used to search out a mottle on the wall of a test pit.

Case History 3 In appealing the decision cited in Case History 1, a formal hearing was scheduled by the Township authorities (a stipulation of the law). The points that were presented at the hearing that were of a technological nature were all framed around the proposition that isolated mottles (or apparent mottles) should not be used as a basis for permit rejection. The specific points made are summarized as follows:

1. Apparent mottles can be caused by a weathering phenomenon called varigation, which also defines a variable color pattern, and is not associated with a wet condition in the soil (i.e., a high seasonal water table).
2. Apparent mottles can be caused by root staining associated with the action of acid rain. (Pennsylvania is documented as one of the states that is, and has been, subject to the highest concentrations of acid rain.)
3. In areas of glacial till deposits, there occur countless small perched water tables. As the name implies, these are pools of subsurface water that are "suspended" in the soil profile in regions above the main regional water table, where the water accumulates in cuplike lenses of clay. In such cases, real mottles of an isolated nature can form. Students of the geosciences can attest to the fact that most of the upper third of the United States contains extensive deposits of glacial till. (Almost all geology texts will show a map of the extent of the most recent periods of glaciation: about 18,000 years ago.)
4. In regions of glaciation where glacial tills predominate, there is a high content of clay in the soil profile. In such soils, water will rise above the water table by capillary action, thus alternately wetting and drying a zone considerably above the water table, causing mottling to develop well above that level.

A conversation I had with a local regulator is also revealing, in which I posed a rather obvious question about how there could be such a major discrepancy in assessment of depth to mottling between the soil scientist/consultant and the regulators. The answer: "I've had instances where two or three soil scientists had different opinions." Clearly, something is seriously wrong with both the written regulations and the manner in which they are enforced. While much of the foregoing is anecdotal in nature, another comment by the former regulator was much broader in scope, and supportive of the argument against the existing situation. He stated that "most local regulators agree that they are not really qualified to judge mottling." He also stated that local regulators had been requesting that they be relieved of this responsibility "for years" at seminars, but that the state authorities "will not budge."

With all of the foregoing as background, I would make the following recommendations:

1. That the language of Section 73.1 regarding mottling be changed to ensure that reasonable agreement among qualified experts will occur.
2. That local regulators be relieved of the responsibility of assessing mottling (and certain other aspects of "permitting").
3. That a new category be established called "Doubtful Sites."
4. That Doubtful Sites be assigned to qualified and certified professionals for assessment of suitability for sewage disposal systems.

The last recommendation will mean that state authorities will have to get into the process of certifying professionals for that purpose, but that effort should not create problems comparable to those that now exist, and would likely be merely a matter of compiling a list of academic and other criteria for applicants, much as is commonly done for the certification of other professionals.

This approach would, I think, have many advantages. If one accepts the proposition that anything that affects design *is* design, then professionals should make the decisions, and whether or not the site in question is suitable for a sewage disposal system would certainly qualify as a *major* design decision. It is, in fact, almost ludicrous that such a decision is presently entrusted to local authorities, almost all of whom are not professionals, and thus not qualified to make decisions regarding design. Secondly, the state would have a highly qualified group to develop better written regulations, thus relieving the state of the sole responsibility for this activity. Since state regulators regularly assert that they are woefully understaffed, such procedures would seem to be beneficial in many respects and for many who would be affected: the state regulators, the local regulators, and, last but certainly not least, the public and the environment. Undoubtedly, the cost of site assessments would be much higher when done by professionals, but I would expect that a person with a designated Doubtful Site would much prefer to have a professional investigate and evaluate the site, even at substantially higher cost. Furthermore, I believe that the tax-paying public is entitled to such an option.

Another weakness of the regulations is the omission of any criteria that recognize differences between usages, as such usages would affect the potential for pollution. A comparison of two types of rural residential developments will illustrate this point. In the Poconos, where tourism and recreation are major components of the economy, there are many developments with a wide variety of usages, Some are of the kind that might be called "vacation oriented." Before 1975, which is about the time that septic systems became a major environmental concern, many such developments were built with little attention to site density (i.e., lot sizes) or septic regulation enforcement. These included many vacation amenities, such as swimming pools, beaches, and sailing facilities. In addition, many homes were built as "TWORPs" (tax write-off rental properties), and were rented throughout the year to the sailing and skiing crowd. The word crowd is here chosen quite consciously, because that is what one often got!

On the other hand, there are the "quiet" developments. The development described earlier (Case Histories 1–3) is an example. It is known as Wagner Forest Park, and has as part of its highway sign the subtitle, "A Conservation Oriented Community." The developer is Sterling Wagner, a highly respected octogenarian whose family has lived in the area for decades. He has a BS and MLE (Master of Landscape Engineering) from Syracuse University, classes of 1927, and 1930. The development was started in 1970, long before there were any laws regarding density. He consciously chose to subdivide the land into lots of a minimum size of two acres (as contrasted to the current law requiring one-half acre minimums in the region), consistent with his desire to develop a "conservation-oriented community." Accordingly, there are almost no amenities. While there is a lake, the people in the community association even voted down a fully sanded beach area that had been considered for swimming. As might be surmised, rentals are rare, and so the attendant potential abuse of septic systems are far less than those associated with vacation developments.

The contrast between these two extreme types of developments could not be more apparent, and yet no provision is made in the current septic system regulations that recognizes such variations. While it would not be practical to write a set of regulations for every development and its individual characteristics, certainly, with some carefully chosen wording, it would be feasible to create a regulation that would allow for major differences such as those cited in the comparison.

9.2.1 Wetlands

In more recent years, a new and controversial subject has evolved: the protection of wetlands, a matter that is intrinsically connected to septic system regulations in particular, and land usage in the broader sense. As most readers will be aware, the national byword, as uttered by President Bush in the 1988 campaign is "No net loss." It sounds simple enough, but the spin-off and ramifications of this simple statement have been dramatic and far-reaching, particularly as a result of still more problems with regulations and the regulatory process.

It is beyond the scope of this book to include all of the pros and cons of this controversial matter. Indeed, things are still very much up in the air as the political campaign unfolds. Much will depend on the "top-down" direction that will come about from the federal administration. However, I would like to offer some observations about the controversies that are currently swirling with respect to the unavoidable impact that wetlands regulations will have on septic system technology, particularly site investigation and evaluation. As a compromise between a full treatment of the subject (which is beyond the focus of this book, and which I, quite frankly, lack the expertise to accomplish, anyway), I will offer only limited commentary, some of which is editorial in nature and deals with an attempt to identify what I view to be some major problems of the process. I will then close this section with a list of references and sources that will enable the reader to become more familiar with wetlands protection, and all that it entails.

DEFINITION AND DELINEATION Before drawing lines on a map of a study site that delineate the presence of wetlands, there must be an agreement as to what a wetland is. There is some general agreement among scientists regarding the parameters that should be involved, but there is as yet no agreement as to what numbers should be used that will reduce the definition to actual lines on a map. The physical parameters are vegetation, soil, and hydrology (water . . . its location and duration, surface and subsurface). A formal definition of wetlands, as promulgated by the Corps of Engineers, is "those areas that are inundated or saturated by surface or groundwater at a sufficient frequency and duration sufficient to support, and that under normal circumstances do support, a prevalence of vegetation typically adapted for life in saturated soil conditions. Wetlands generally include swamps, marshes, bogs, and similar areas."

The problems arise in getting agreement among interested parties on quantification of this definition. The first attempt, that currently in effect, is that land must be saturated within 18 in. of the surface at least seven days during the growing season to be considered a wetland. However, William Reilly, current federal Environmental Protection Agency chief, claims that the current wetlands definition is unworkable and causing too much confusion among property owners. Accordingly, a new proposal is that a wetland must have standing water for 15 days or be saturated to the surface for 21 consecutive days. This proposal would shrink the protected land by significant amounts.

As I write, the battle rages between the land users (developers, lot owners, farmers, miners, etc.) and the environmentalists. Obviously, this is a very political matter that could, and probably will, swing back and forth as administrations change at all levels of government.

BUFFERS Once a wetland is defined and then delineated, the question arises as to its protection against degradation by adjacent activities of all sorts. To prevent or minimize such pollution, the notion of buffer zones (also called transitional areas) has become prominent in wetland protection regulation. As might be surmised, this is, if anything, even more controversial than wetland definition criteria. Numbers currently on the books in various jurisdictions (e.g., New Jersey and Pennsylvania) range from 150 to 300 ft as specified buffer dimensions. The size of the buffer zone and the activities permitted in the buffer zones (if any) depend on both the value of the wetland and the nature of the proposed activity in the buffer zone.

WETLAND CATEGORIES Depending on the sensitivity and importance of the wetland, various jurisdictions typically establish a number of categories of wetland. In Pennsylvania, for instance, there are currently three categories: Exceptional Value wetlands, Important wetlands, and Limited Value wetlands. Attempts have been made, with mixed success, to develop methods of quantifying the required size of a buffer zone for a given set of conditions. For example, Roman and Good, of Rutgers University, developed a wetland buffer model for the Pinelands of New Jersey to arrive mathematically at a buffer width ("Wetlands: Untangling the Issues and Regulations," 1989, p. 33).

However, as Gross points out, "[T]he fallibility of that model is that [they] started out with the assumption that 300 feet was an appropriate wetlands buffer and what the model does is reduce the buffer depending upon various circumstances." ("Wetlands . . .," 1989, p. 47).

In a course in high school chemistry, I recall doing a laboratory experiment that passed contaminated water through a column of a granular material of a resinlike character (zeolite, I believe it was called), and I was impressed, even awed, by the fact that the water was totally "purified" by that simple desktop process. As is well known, there exist today a variety of sinktop commercial water purifiers that accomplish similar miracles. On a much larger scale, water purification plants use sand and activated charcoal as a filtration medium, usually of perhaps a couple of feet in thickness. As was noted earlier, natural soils of a silty texture perform the job of purification in septic system design, wherein a 20-in. thickness is required for approval of a sand mound system, and a 60-in. thickness is required for a "standard" system.

My question is this: Since a buffer zone involves essentially a filtration medium for the horizontal movement of water toward the wetland, what is the justification for going from inches (for vertical filtration requirements) to hundreds of feet for buffer zones? As far as I know, no one has ever raised this question. It seems to me that this would be a better place to start the evaluation of buffer zone widths, rather than just picking numbers out of thin air, as so far seems to have been the case. (As Gross pointed out, "[T]he New Jersey legisla ture in its infinite wisdom determined that the maximum buffer that you could impose is 150 feet. I have no idea where they got that from except one could assume that they took the 300 feet that is used by the Pinelands Commission [also apparently chosen arbitrarily] and divided by two" ("Wetlands . . .," 1989, p. 48).

COMMENTARY With the foregoing as a brief summary overview (about as brief as this loquacious Irishman ever gets), I would like to offer some commentary, much of it editorial, about some of the problems with wetlands policies as they now exist, and the directions in which they may develop.

Wetlands policy is in a state of infancy and flux in almost all jurisdictions. From what I have been able to glean from my research, New Jersey is somewhat further along than other states, and yet legislation to regulate freshwater wetlands was not introduced until 1983. Four years of study and negotiations followed, with further delays for adoption. Thus, it was about 1989 before enforcement was to begin. It is perhaps understandable that New Jersey would be in the forefront of this effort, for a number of reasons. It has a number of large and very sensitive freshwater wetlands, from north to south: the Hackensack Meadowlands, the Great Swamp, and the Pinelands, to mention a few. Also, these areas are surrounded by some of the most densely populated and/or industrialized areas in the United States. Thus, pressures for development are substantial.

Complicating the process greatly is the state of flux at the federal level, exacerbated even further by the extremely political nature of the controversies, notably between the environmentalists and the developers.

My perception of the current state of affairs is that there is too much "micro-regulating" being attempted, so much so as to make the system "unworkable" (a word used by William Reilly, federal EPA chief, in Fall 1992, regarding the current wetlands definition). Some case histories and quotations will illustrate what I mean by micro-regulating.

I recently talked with a farmer from northern Pennsylvania, near Towanda, who was more than a little upset over new wetlands policies being enforced by the DER that identified wetlands (he gestured with his arms) about the size of a manhole. He was told that he could not plow within 200 ft! A newly proposed definition at the federal level rejects the theory that "every mud puddle is a wetland" (*Pocono Record*, 10 August 1991). Other farmers have protested that ditches that need to be cleaned out are now a "protected resource," and that engineering studies are now needed to clean out a farm pond. The farmer with the manhole-size wetlands also told me that he is no longer permitted to drive cattle across a creek; that bridges will have to be constructed for such a purpose!

An example of micro-regulating in the area of home construction is afforded by the not uncommon situation in which a person purchases a lot that is surrounded by lots with existing homes, only to discover that an inspection by local regulators reveals that a portion of the lot is declared wetland, thus often generating extensive amounts of further site work and related hearings, and legal work (and fees!) that may or may not result in the new house being approved. In many such cases, the size of the declared wetland is quite small, in the tenth of an acre range or even less (as in our farmer friend's plight).

As noted in the previous section, one of the parameters currently used to define wetlands is vegetation. At a wetlands conference in 1989, a scientist asserted that "a wetland field survey by a botanist is required to determine the dominant vegetation on a site" (Collier, 1989). If every quarter-acre site in America has to be assiduously searched for the elusive plant species and soil mottle, there are not enough botanists and soil scientists in the world to accomplish the task, and there never will be. Even if there were, the collective cost of their professional services would be prohibitive, and with the sluggish state of the national economy, and with many other pressing needs such as affordable housing, infrastructure needs, health and education crises, and problems of the inner cities, our society can ill afford to spend so much on the micro-regulation of wetlands that has been described. Moreover, there is a way to accomplish much more freshwater wetlands protection at a tiny fraction of the cost of micro-regulation, namely through vigorous efforts toward ensuring that septic systems are properly constructed and, most especially, maintained effectively. I shall address these matters in subsequent sections.

What is needed, then, is a realistic approach to wetlands definition and a recognition of the fact that the sensible approach would and must be to ensure that *obvious* wetlands of substantial size are not abused. This would be doable, and would not result in the enormous amount of wasted energy that now goes into an impossible task of trying to save every mudhole. As Dean Kleckner states, "We have argued all along that lands classified as wetlands should be truly wet" (*Pocono Record*, 10 August 1991). Exclusions based on small size (manhole-size "wetlands") or alternative uses (farm ponds) should also be incorporated into new definitions.

Finally, before getting back to the more mundane, technical aspects of site investigation, I would like to address a matter that has concerned me for some time, namely an attitude about site evaluation on the part of many environmentalists and particularly those regulators who interact on a daily basis with the public. There were many genuine and even blatant abuses of the environment that took place prior to about 1973, when nobody was looking. Greedy and unscrupulous developers filled in swamps at will, created tiny lots, and duped people into buying the lots at inflated prices: take the money and run. I believe that the present attitudes of regulators is largely based on an overreaction to the abuses, and that many regulators now regard anyone who wants to build anything as being a despoiler of the environment and that they see their mission to be that of avenging angels who will prevent the despolation.

I know of a case where a DER staff person arrived on a site (with three colleagues "for support"), looked out at the site, and said in a distinctly audible voice, "You can't build here!" (Case History 1, described earlier). This was said *before* any inspection of the test pit was done! Upon inspection of the test pit, it was declared that mottling occurred at a depth of 10 in., as noted earlier, contrary to the opinion of a soil scientist whose opinion was overruled.

I believe such zealotry is prevalent at lower levels of the regulatory bureaucracy partly because they "get the message" from the top, and are fearful that their job performance will suffer if they don't crack down on the developers and lot owners. Add to that the badly conceived and carelessly written regulations, and the situation becomes, as Reilly said, "unworkable," not to mention extremely expensive and aggravating to a large number of very responsible people who try to deal with the mess.

If I appear to exaggerate, consider this: On Tuesday, 29 September 1992, there appeared in the *Pocono Record* a story about the passing of a prominent and highly respected developer, Russell Altemose, who was said to have "changed the face" of the West End. Mr. Altemose had succumbed after a series of heart attacks. According to the article, "Altemose blamed his first heart attack . . . on a dispute over septic system regulations with the state Department of Environmental Resources officials."

PERCOLATION TESTING The third phase of the site investigation is a series of percolation tests, which are performed at the proposed location of the drain field. These are undertaken after (and *only* after!) the site has been evaluated for wetlands, and has passed the probe inspection for suitability for mound or standard septic construction (i.e., 20 in. of suitable soil for mound construction, and 60 in. for standard septic).

As the name connotes, the test is done to establish a numerical value that represents how the site will drain when effluent from the house is discharged to the septic tank and then to the drain field. In Pennsylvania, the specifications for the test are described in Section 73.15 of Chapter 73, *Standards for Sewage Facilities*. The essentials of the test program involve the preparation of six or more holes, which are then tested by filling them with water and observing the rate at which the water percolates into the surrounding and subsurface soil. Following are pertinent excerpts from the descriptions of the test procedures.

> 73.15 (3) Type of hole. Holes having a uniform diameter of six to ten inches shall be bored or dug as follows . . .
>
> 73.15 (4) Preparation. The bottom and sides of the hole shall be scarified with a knife blade or sharp-pointed instrument in order to completely remove any smeared soil surfaces and to provide a natural soil interface into which water may percolate. All loose material shall be removed from the hole. Two inches of coarse sand or fine gravel shall be placed in the bottom of the hole to protect the soil from scouring and clogging of the pores.

The remaining steps deal with presoaking the hole and taking measurements of the fall of water in the hole. The manner in which the holes are filled with water is not specified other than to instruct, "holes shall be filled with water," and "water shall be placed in the hole."

I believe there are deficiencies in these instructions that can result in very substantial errors in the "perk rate" obtained. I have run side-by-side tests, following the above instructions on one hole (which I shall call PA/standard), and using modified instructions on the other (which I shall call Winneberger, after the person from whom I learned of the modified techniques). Figure 9.2 shows schematics of the two holes.

The Winneberger hole is bored undersize, and is then enlarged to the PA/standard size by carefully cutting vertical segments with a spatula or knife; the hole is enlarged by gently tearing the soil away from the eventual perimeter face of the enlarged hole. As with the PA/standard hole, the loose soil is carefully removed from the hole. Newspaper is cut to fit the bottom circle (Yankees 12, Mets 6), and additional newspaper is rolled into a loose cylinder to fit the hole. (Holes may be poked in the newspaper beforehand with a pen or pencil, but Winneberger commented that this refinement is probably not necessary for free drainage.) Water is then placed in the hole via a tube connected from a suitable container and extended to the bottom of the hole. With a suitable pinch clamp, the water is then allowed to enter the hole *with no free-fall*, and in a careful manner. With very little practice or extra effort, the hole will be filled with water so clear that the newspaper will be easily readable.

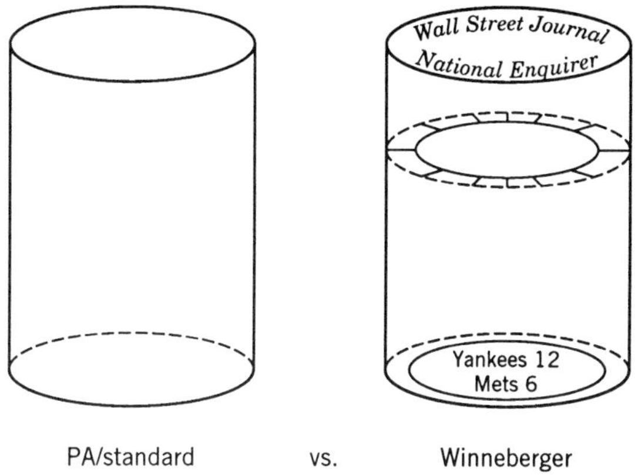

Figure 9.2. *Percolation test, illustrating the efficacy of trench systems over bed systems.*

When I saw these side-by-side test results, and later duplicated them myself, I was, quite frankly, amazed at the difference: The perk rates for the Winneberger hole were as much as ten times higher than that for the PA/standard hole. (Readers may have to do the test for themselves to believe it, I suspect!) Some thought will reveal the reasons for this startling difference. First, there is little doubt that the Winneberger method of preparing the hole will result in a sidewall that is virtually 100% undisturbed, i.e., completely free from smearing. This would not likely be so for the PA/standard hole where instructions merely call for scarification with a knife blade. Also, unlike the PA/standard hole, there would be little or no surface compaction in the Winneberger Hole.

The second feature, relating to the introduction of water into the respective holes, is simply that the Winneberger method is far superior to the PA/standard in ensuring that there will be little or no "slurry creation" caused by erosion of soil upon introduction of water into the hole. The instructions for the PA/standard hole ("holes shall be filled with water") would, almost certainly result in the hole being filled with turbid or even opaque slurry, rather than clear water. Absent the explicit instructions of the Winneberger method, the technician would simply pour water into the hole from some height that would generate erosion on impact. Thus, in a PA/standard hole, one would start with a (partially) smeared wall that is also substantially compacted (by boring), and which will be further clogged by the solid particles in suspension in the slurry as seepage occurs laterally through the sidewalls. The result would be a predictably lower perk rate. As noted, the magnitude of the difference is what was unexpected.

The ramifications of all of this are directly related to design and construction.

9.3 DESIGN AND CONSTRUCTION

The first step in the design phase is to select a system that is "allowable" within the existing parameters set by the appropriate governing bodies, applicable laws, and regulations. The choice is specifically governed by the results of the site investigation and evaluation: wetlands evaluation, probe inspection, and perk test results.

For opurposes of the following section, it is assumed that the site has been deemed "viable" (or buildable) using either standard septic (sometimes also called an "in-ground" system) or for a mound system (also called "sand mounds," or "turkey mounds"). Thus, for the site in question, wetlands are not present, the probe inspection reveals no limiting zone shallower than 20 in., and the perk rate is acceptable according to the stipulated regulations for standard or mound construction. After describing aspects of the design and construction of these common systems, a brief summary of alternatives for problem sites will be presented. A bibliography and addresses will be provided that will enable further study of a system that may be of interest.

I limit my commentary throughout to those aspects related to points that I have presented in previous sections, for two reasons. First, my specialty in civil engineering is geotechnical. Thus, while somewhat knowledgeable (by self-study) in the area of environmental/sanitary engineering, I do not have the detailed, comprehensive expertise and experience necessary to deal exhaustively with design and construction of septic systems, but I do feel that I can contribute effectively in those areas relating to soils and seepage. Also, I believe I can contribute suggestions for the attention of those who *are* expert in design and construction of septic systems. Second, two books have recently been published that deal with septic design and construction in detail (Perkins, 1989; and Kaplan, 1991), both of which added considerably to my knowledge about septic systems. The books are recommended to those with special interests in the field.

9.3.1 Standard Systems

Figures 9.1*a* and *b* show two standard ("in-ground") systems that are identified, respectively, as "Trench" and "Bed" systems. I have reproduced them exactly as they appear in the EPA document in which they appear, including the verbal descriptions. The reader will note the one addition made by me: the notation "sic", following the assertion that trench systems are the "most common on-site system." *I do not believe that this is anywhere near an accurate statement.* I write this based on my observations and conversations with contractors of long experience in the Pocono region of Pennsylvania. In 23 years, I have never seen a trench system being built; all were either bed systems or (in recent years) mound systems (Figure 9.1*c*). A contractor of my acquaintance, with perhaps 40 years of experience that included building septic systems, told me that he once had a problem job where he convinced the local SEO to let him try an experimental approach. His intuition was that a series of trenches would work better. *It did*! The contractor did not know why, but I would submit that the reasons are as follows:

1. Sediment accumulates only on horizontal surfaces, a rather obvious statement, but nonetheless true. It follows that, in time, the horizontal surfaces in a drain field will become less permeable, the progression depending to a large extent on the nature and amount of suspended solids that enter the drain field from the septic tank. (Suspended solids are those of a very small size that do not settle rapidly and harmlessly to the bottom of the septic tank, but instead remain in suspension for extended periods until such time as they flow out of the tank to the drain field during periods of large fluid discharge, where the solids eventually clog the drain field. High discharge periods could come about through house parties, extensive use of washing machines and showers, and so forth; more on this in the section on maintenance.)

   In addition to simple mechanical sedimentation of inert suspended solids, as Perkins (1989) explains, a "biological mat is formed, which is beneficial in filtering out harmful bacteria that are contained in the effluent stream." Obviously, if this mat becomes clogged by the inert suspended solids, the horizontal surfaces of the field will cease to function effectively, and may eventually fail.
2. When horizontal surfaces become less effective, sidewall drainage will take over. Trench systems have a higher percentage of sidewall area than do bed systems.
3. The horizontal permeability of natural soils is much greater than vertical permeability, because of the natural stratification that takes place when soils are deposited geologically. Casagrande (1940), in his definitive work, "Seepage Through Dams," states that the ratio is at least 10, and may be as much as 100 for the more distinctly stratified soils common in sedimentary deposits.

All of the above explains why the Winneberger perk test yields such dramatically higher perk rates than does the PA/standard test. The superiority of the trench system over the bed system should now be evident.

Unless the Pocono area is an aberration with respect to septic system construction, and I have no reason to believe that it is, bed systems are the norm across the United States. Maybe the "feds" know something I don't, but I think it more likely that they merely think that trenches are most common. In any event, they should be, and I would recommend that consideration be given to making the design mandatory, particularly if regulatory bodies are serious about minimizing pollution of groundwater.

With regard to construction, I believe bed systems are more common simply because they are easier to construct, since large trucks and other construction machinery can easily negotiate an open bed. Also, since the amount of excavation would be greater for a bed than for a series of trenches, payment to the contractor would be correspondingly higher, since payment for excavation is commonly on a per yard basis. Thus, the contractor has a number of economic incentives to opt for the bed system, given a choice.

Since the machinery is heavy, this creates another problem: the likely sealing of large portions of the drain field by unavoidable compaction by construction traffic. Thus, in bed systems, the drain field is partially clogged before the first toilet is flushed! In trench system construction, this problem is automatically diminished, since the heavier equipment (i.e., trucks) cannot enter the trenches. In trench construction, of course, the work is more hand-labor intensive, but I view that as a positive feature, since careful attention and care can be given to trench preparation.

TRENCH PREPARATION AND CONSTRUCTION An important detail of trench preparation, in light of the importance of sidewall drainage, is the careful (even assiduous!) attention that should be given to sidewall exposure. I would recommend hand labor with a pitchfork or steel rake to pull off segments of the sidewall after the trench has been excavated by a backhoe. Careful cleaning of the bottom of the trench should proceed as the work is done. As a final step, it is recommended that the bottom of the trench be gently raked to loosen areas that have been compacted by footprints. The soil thus loosened should then be removed with a suitable heavy-duty vacuum. This process is done to simulate the preparation of the Winneberger test hole done for the percolation testing, the intent being to create sidewalls and bottoms that are as undisturbed and smear-free as is reasonably possible.

Perkins recommends (I was pleased to see!) a backhoe bucket with side teeth to rough up the sides of a trench (Perkins, 1989, p. 78). I think the pitchfork or rake would do a better job, because of the almost sculpting nature of the hand labor, but the idea is the same. While some might disagree, I believe that a septic system that is properly designed, constructed, and maintained for a site that has been correctly investigated and evaluated can have an indefinite service life without polluting the groundwater. Thus, a few extra hours of careful work on trench preparation is, I believe, a good investment.

I would also strongly recommend the use of filter fabric in trench construction (Perkins, 1989, p. 73), the purpose of which is to protect gravel backfill (if used) from intrusion by finer-grained soil particles. Inspection and monitoring of the construction process is a must. I think that here is a function that SEOs could perform very well. The SEO should be present for at least the most important parts of the trench preparation, especially to instruct the start of the sculpting process of the sidewalls, to ensure that the contractor and the workers are doing the work assiduously. I would expect that contractors and workers would be cooperative if it is explained briefly to them how important such seemingly innocuous details really are toward the long-term effectiveness of the system. And owners, I think, would be happy to pay for the extra effort that would be involved. It may very well be, moreover, that the total cost of trench construction may be less than bed construction, since the unit costs of the work would be much lower than for the heavy equipment and heavy equipment operators that are commonly used for bed construction.

Finally, I might even recommend to the homeowners who might read this book that they do a little inspecting and monitoring of their own, inasmuch as the replacement of a failed septic system might run into five figures in 1993 dollars.

Trench Backfill Historically, the usual method of construction involves the use of washed gravel as backfill, wherein about one foot of gravel is first placed to the correct level and slope to provide the support for a 4-in. perforated pipe. (These are the pipes shown emanating from the distribution box in Figure 9.1*a*, the levels and slope of which are dictated by hydraulic considerations of the design.) Additional gravel is then placed around and above the pipe, followed by a covering and, lastly, a backfill of (usually) native soil to grade. The covering above the gravel is intended to prevent soil intrusion into the gravel void spaces. For similar reasons, as suggested earlier, the gravel fill should be encased with geotextile (filter) fabric.

Alternative methods of trench backfill require little or no gravel, such as with the use of precast concrete leaching chambers. These have been used with excellent results, but require heavy lifting equipment because of their great weight, so costs are considerable.

A new and apparently very attractive alternative is afforded by an Infiltrator leaching system (May, 1989). This system has as its principal feature inverted U-shaped elements called Infiltrators that are linked together and placed in the prepared trench to form a tunnel. The units are 3 ft wide, about 6 ft long, and 1 ft high, made of molded polyethylene, and weigh only 25 lb. The sides of the units incorporate micro-leaching chambers (a series of horizontal slots) that (they claim) eliminate the need for geotextiles, and the tops are solid and thus prevent any soil backfill intrusion from above.

(The manufacturer recommends coarse–medium sand as backfill at the sides of the units between the units and the trench wall. As a soils engineer, I would concur. I would also suggest that an appropriate filter fabric, attached to the slotted openings of the units, might be considered as additional protection against the possibility of lateral intrusion of soil through the slots of the unit. If fabric is chosen, it may be possible to eliminate the select backfill in favor of the cheaper natural soil.)

Other advantages claimed by the manufacturer/inventors are impressive. Indeed, most of them seem to me quite logical and incontestable. The principal advantage is the virtual elimination of masking. When gravel is used, the areas of interface between the stone particles and the prepared surfaces of the trench (called infiltrative surfaces by the manufacturers) are obviously not available for percolation of effluent. This masking of the infiltrative surface is estimated by independent investigators to be about 50%, which seems reasonable to me. Because the space under the U-shaped element is empty, the entire bottom of the trench is unmasked. As noted, the sides of the elements are claimed to minimize the masking effect.

Other major advantages accrue because of the light weight of the units, including ease of delivery and installation, and reduced labor and equipment costs. Another significant advantage is ease of inspection and correction (if ever needed), especially when one compares these features to the difficulties inherent with gravel systems. Indeed, the cost of the repair or renovation of marginally performing gravel systems would, I think, approach that of total replacement, one of the real deficiencies of the system. In a word, if it doesn't work, you're in trouble! (More on this in a subsequent section.)

The bottom line is the claim by the manufacturer that trench length can be reduced by as much as 50%. The manufacturer reports that the system has been approved by the U. S. Department of Housing and Urban Development–Federal Housing Authority.

Lastly, what impresses me is the fact that the manufacturer indicates that the company has provided investigators at the University of Wisconsin with the components of the system "to do comparison testing of the unit, next to conventional systems in differing soil groups" (May, 1989). Investigators at the University of Wisconsin are some of the most active and respected researchers in the United States. They have been pursuing applied and full-scale research into septic system technology for many years. Their publications are highly recommended.

9.3.2 Mound Systems

Figure 9.1*c* shows the cross section of a mound system, adapted from the EPA drawings cited previously. Proceeding from the top down, the elements of the mound system are as follows:

1. *Seeded topsoil.* The grass cover provides a means of augmenting the system's water capacity by allowing for the removal of water by evapotranspiration, and also diminishes erosion. Additionally, the grass cover serves as an aesthetic feature. However, trees and large plants should be avoided, because root systems could develop that would be detrimental to the proper functioning of the system.
2. *Hay–clay–hay strata.* The clay is a relatively impervious cap to prevent large amounts of rainwater from entering the system. The hay prevents the infiltration of clay into the crushed stone.
3. *Crushed stone.* The washed gravel upon which the perforated pipe distribution system is laid. The perforated pipe is, in turn, connected to the pump chamber. Effluent from the septic tank flows by gravity to the pump chamber (not shown). The pump is automatically programmed to be activated to pump the effluent upward to the distribution system of perforated pipes. Infiltrators can be used in place of the crushed stone, according to the manufacturer.

4. *Sand.* This select fill is imported to the site, and is intended to provide the necessary cleansing action to the effluent as it percolates downward into the natural soil. As noted earlier, the mound system is required when there is an inadequate depth of suitable natural soil to the limiting zone. In Pennsylvania, this depth is 20 in. The texture of the sand should be fine enough to impart marginal draining characteristics to the soil. (See Section 6.2.1 for a description of marginally permeable soils.)

ASPECTS OF DESIGN, CONSTRUCTION, AND MATERIAL SELECTION The term "aspects" was consciously chosen as the heading of this section because of my conviction that some very important matters need to be addressed with respect to mounds, the most important of which is the use of an adequate soil classification system for material selection and specifications, and, for that matter, all other facets of earthwork related to septic system technology. The scope of this chapter does not permit a full treatment of all design and construction aspects, and so I will limit my comments to those that I think need serious attention and change. To allow readers to learn about all features of the art, I strongly recommend an exquisitely detailed paper by researchers at the University of Wisconsin, in which no less than 20 steps are described for the construction of the mound, along with all other details of design and construction of the system (Converse et al., 1975).

In the mound recommended by the Wisconsin researchers, there is included a trench or trenches at the top of the cross section of select fill, consisting of ⅝ in.–1 in. stone (in place of the crushed stone depicted in Figure 9.1*c*). The size, number, and geometry of the trenches is dictated generally by the number of bedrooms in the home. The perforated PVC pipe, serving as the effluent distribution system, would rest on the surface of these trenches. Wisconsin researchers, and other professionals in the field, recommend a "pressure distribution system" to carry the effluent through the trenches and downward through the select fill to the subsoil below. Pressure distribution systems provide (theoretically) an even distribution of the effluent over the drain field area, as opposed to gravity flow systems, wherein the effluent is often concentrated inefficiently over smaller areas. Details of the design of the "plumbing" necessary to achieve equal distribution of effluent are contained in the Wisconsin paper, and in other sources cited herein.

Select Fill: Soil Classification Systems In the spring of 1993, I attended a three-day seminar at Cook County College, the Extension Division of Rutgers University, the subject of which was Septic System Technology. A significant number of professionals attending expressed the opinion that select fills were a major problem with mound system performance. Interestingly, no one seemed to know why. I think I do. In fact, I'm virtually certain I do. It is the use of an inadequate soil classification system in septic system technology, namely the USDA system (United States Department of Agriculture). Logically enough, this system was developed for identifying and classifying soils for agricultural purposes, and is a finc system for that purpose. Soils engineers, recognizing the need for a system more suitable for engineering purposes, have developed their own systems. Two of the better known are the AASHO system (American Association of State Highway Officials . . . now known as AASHTO, as a result of greater emphases on transportation matters), and The Unified Soil Classification system, the latter being an attempt among engineers to develop one system that would be universally accepted . . . at least among engineers.

To muddy the waters further (pun intended), I am a proponent of still another system, one that is commonly used by those of us in soils engineering who were influenced heavily by its developer, Donald Burmister, of Columbia University, one of the early and highly respected pioneers of the then new discipline of soil mechanics. Appropriately enough, it is called the Burmister system. I believe it is accurate to say that septic technology has advanced to the point where it is a lot closer to engineering than to agriculture, and so an engineering classification system would better serve all aspects of the art. In earlier sections, I have noted that soils of a silty nature are best suited for the cleansing action that occurs as septic effluent passes through the soil. But such an assertion is inadequate when it comes to actually choosing or evaluating a potential select fill to be used in a mound. There must be ways to identify and classify a soil to some reasonable degree of specificity, and I believe the Burmister system does this best.

Again, the scope of this chapter does not permit a full treatment of soil classification systems. Instead, I provide some limited commentary as to why the Burmister system is superior, and what I perceive are some of the deficiencies of the other systems that create problems, the problems I have long suspected, and which were confirmed by comments of participants at the Cook College seminar.

At that seminar, a speaker addressed the subject of the appropriate soil texture for select fills for mounds. Virtually no distinction was made between silts and clays, and no distinction at all was made relating to variations of properties among different clays. The speaker was not a soils specialist, but was extremely knowledgeable about septic system design details dealing with the hydraulics, the hardware design, and the sanitary engineering aspects of septic technology (what I called earlier "pollution science"). Not wishing to embarrass a fellow professional, I did not press the matter from the floor, but I made a note that I would address this problem here. And so I do. (Parenthetically, I have come to perceive that septic system technology is extremely complex, requiring an unusual number of "crossover" skills, everything from wetlands ecology, sanitary engineering, geology, and soilis engineering. *Whew*!)

In the Unified system, ostensibly intended to be an improvement over the AASHO system—and I believe it is—there is still a deficiency regarding proportions of constituent soil types, of which there are four recognized categories: gravel, sand, silt, and clay (in descending order of size). If one examines the detailed rules of the Unified system, one will detect that one need simply decide the relative amounts of constituents present in the sample. Thus, if a soil contains only sand and gravel, you must merely decide which constituent is present in the greater amount. If sand, then the soil is classified simply as a "gravelly sand" (an SG, in symbols); if gravel, it's a "sandy gravel" (GS). Thus, a soil that has 99% gravel has the same classification as one that has 51% gravel. If the soil has all four constituents present, it may be a silty, sandy, clayey gravel, and all one gleans from the description is the relative order of constituents present in the soil, i.e., gravel most, silt least. But note that this could be 1% silt, 2% sand, 3% clay, and 94% gravel. It could also mean 23% silt, 24% sand, 25% clay, and 28% gravel . . . or any combination of percentages between these proportions! Both of these hypothetical soils would have the same Unified description, but the engineering behavior of these soils would be widely different.

On this basis, I would contend that there is something wrong with the classification system. A good system is one that forces the classifier to use rules that will result in the soil being classified within a reasonably close range of engineering *behaviors.* The Burmister system does this, and does so in a rather straightforward and logical manner. It is also a very practical system, in that it permits the classifier to quickly determine the classification by simple visual and tactile means *in the field*. Following is a brief description of the essential features of the system.

As with other engineering classification systems, there are four recognized constituents: gravel, sand, silt, and clay, in decreasing order of sizes, respectively. Gravel, sand, and silt are defined on the basis of size, and each is further subdivided into three grain size fractions: coarse, medium, and fine (c,m,f, in the shorthand notation commonly used in written descriptions). The maximum size (for coarse gravel) is 60 mm, and the descending sizes are arranged in an easily memorized sequence of 60, 20, 6, 2 (c,m,f gravel), 0.6, 0,2,0.06 (c,m,f sand), and 0.02, 0.006, 0.002 (c,m,f silt). If these size boundaries are plotted on six-cycle semilog paper (*descending* sizes left to right), as shown in Figure 9.3, the result is a series of nine vertical bands of equal width, each of which represents the nine grain size fractions. It is useful to those learning the system to develop an assortment of devices to help with the visualization of the Burmister sizes. One such device is to relate the sizes to common objects. Thus, a billiard ball (about 57 millimeters) can be envisaged to represent the approximate upper limit of coarse gravel (cG), a golf ball (40 mm, cG), an olive or marble, mG, a raisin, fG, and a BB (as in BB gun), the lower limit of fine Gravel, and, simultaneously, the upper limit of sand sizes. A medium sand (mS) could be represented by granulated sugar or table salt. Fine sands are those soils in which one can just distinguish individual particles with the naked eye; the "fluffier" sands that one finds in the better sand bunkers is a good *very* familiar example of fine sand to my fellow duffers out there.

Another device would be to simply draw circles to true scale for each of the six visible grain fractions, cG to fS. The best device of all would be to prepare

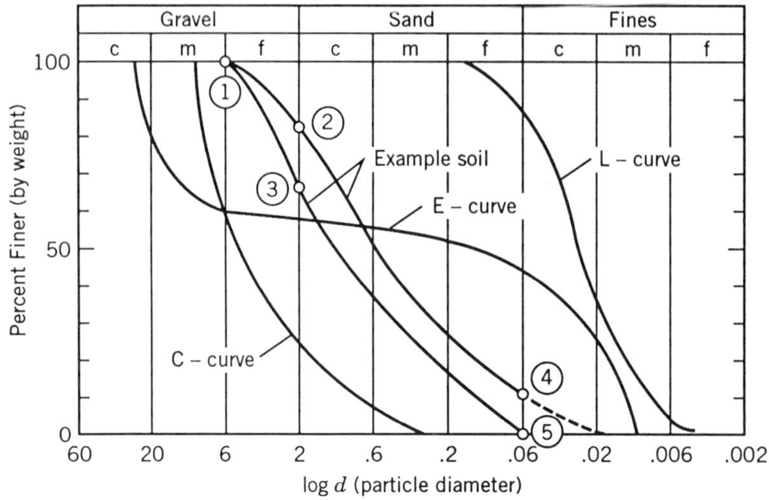

Figure 9.3. *Grain size distribution curves.*

jarsful of real soils of each grain size fraction, using a nested set of the appropriate sieves. (All soils laboratories would have the required sieves.) In this way, a learner could use all senses to practice identifying sizes, i.e., sight and feel. Anything smaller than .06 mm is informally called "fines" by soils engineers, or "minus-200" material. The 200 refers to the sieve size, and its dimensions are in openings per square centimeter. Thus, the larger the number of the sieve, the finer the mesh of the sieve. (The 200 sieve looks like a fine cloth, as contrasted to, say, the No. 10 and the No. 40, where distinct openings are visible. Any soil with particles smaller than 0.002 mm is clay. (It might be noted that nature is infinitely variable, so these boundary definitions are somewhat arbitrary, but are based on the perceived differences in behavior by those expert in the soil sciences and engineering.)

While particle sizes are used to define gravels, sands, and (to a somewhat lesser extent) silts, the important characteristic of clays is the degree of *plasticity*, as defined by the plasticity index (PI), which is the numerical difference between the liquid limit and the plastic limit of the soil. These limits are moisture contents, as determined by standardized laboratory tests, and are done on the "minus 40" fraction of the soil: that portion of the soil that is washed through a No. 40 sieve. (Although theoretically one should use the 200 sieve to separate the fines from the granular particles, research has shown that the inclusion of materials between the 40 and 200 sieves does not significantly affect the results of the limits tests.) The PI, then, is the range of moisture contents over which the soil acts as a plastic material; the higher the PI, the more plastic the soil.

The Burmister system has a very practical feature that enables one to make assessments of plasticity by relatively simple field tests that require visual and tactile senses, plus a certain amount of diligent practice. These tests include dilatancy, kneading, rolling, smearing, and dry strength. An additional test, one of my own invention, is the wash test.

Remembering that the sole purpose of this series of tests is to rate the soil's degree of plasticity, the first test performed should always be the dilatancy test, because a strong positive dilatant reaction is a clear and certain indicator that the soil is a silt, and would be rated NP, or nonplastic. If the test does not produce a strong positive dilatant reaction, then the other tests must be done to rate the degree of plasticity of the clay soil. Because the dilatancy test is so important, I shall here give a fairly complete description of how the test is performed, and then add very brief descriptions of the other follow-up tests.

The Dilatancy Test Mix a portion of the minus-40 material with water to produce a wafer or "pat" of saturated soil about the size of a half-dollar. Arrange the soil in the palm of your hand. Squeeze the wafer by moving your thumb toward your pinky; open the hand to a flat position. Now tap the hand several times with the heel of the other hand. Observe the surface of the soil wafer during repeated squeezing and tapping. If the surface becomes dull upon squeezing and then shiny and wet-looking upon tapping, you have a positive dilatant reaction. Case closed; it's a silt, NP. That's all there is to it. If you don't get a dilatant reaction, the soil possesses some degree of plasticity, and you must then proceed to the other tests.

Kneading From the saturated state, knead the soil as you would pizza dough. The longer it takes to reduce the soil to a crumbly state, the greater the degree of plasticity.

Rolling From the saturated state, roll the soil on a flat, smooth surface into a thread. The ease with which this can be done, and the smaller the thread diameter, the greater the degree of plasticity.

Smearing Form the saturated soil into a ball about the size of a marble. With a knife blade or your fingernail, smear the soil. A shiny or greasy appearance is an indicator of higher degrees of plasticity. A dull appearance indicates lower plasticity.

Dry Strength Allow the ball of soil to air-dry overnight. If the ball crumbles to dust with just a little finger pressure, it is a silt. The higher the dry strength, the higher the plasticity.

Wash Test The more difficult it is to wash the soil from one's hands, the more plastic the soil. (A silt will wash off with gentle shaking underwater.) If, while washing, the soil feels squishy and slimy, and is difficult to remove, then a high degree of plasticity is indicated. Since you have to wash and dry your hands anyway to write down the results (or go to dinner, for that matter), this is a very sensible finishing test. It is also helpful to note that silts will feel gritty, as opposed to squishy and slimy. I once met a scientist from the Soil Conservation Service (SCS) who asserted that he had a colleague who would put a small sample of a soil being evaluated into his mouth, claiming that the gritty texture could more easily be detected in this manner. I would caution against emulating this dedicated soul, especially if your site is in an urban, industrialized area, where all manner of contaminants may be present (radon, lead, gasoline, etc.) not to mention the residue of our four-footed friends.

The Burmister system has a series of terms relating to the results of the field plasticity tests, among which are (for example) silty clay, and clayey silt. As with other systems, the noun predominates, so a silty clay has a significantly higher degree of plasticity than does a clayey silt.

Regarding proportions of the soil constituents, the Burmister system uses common conjunctions, adjectives, and nouns to designate specific percentage ranges, as follows:

> . . ., and means 35–50% . . ., some means 20–35% . . ., little means 10–20% . . ., trace means 0–10% . . . and, means 50–65% . . . some, means 65–80%.

The latter two conventions are not often needed or used. Note that the placement of the comma, either before or after the word, changes its meaning.

Notice that Burmister made the careful judgment that, with some diligence and attention, a person should be able to estimate by inspection, what percentages of a constituent are present in a soil sample to within a degree of accuracy of *no worse* than 15%, and further, that it was reasonable to expect that the gradation of each constituent could be approximated by inspection and feel. No other system makes such demands on the classifier. Can all of this be done? After 26 years of teaching undergraduates the techniques, I can safely assure the reader that the answer is a resounding yes. A considerable amount of diligent practice will be necessary to develop the skill. In time, it is possible to examine a soil for perhaps no more than 20–30 seconds before writing down a fairly accurate description; initial attempts might take one-half hour or more, and be quite inaccurate.

A typical Burmister description is as follows:

> Dense brown c-f SAND, some (-) f Gravel, trace Silt; rounded Gravel, faint odor of gasoline (?), occ. to frequent rounded boulders in field.

Here are some additional rules of the system:

1. The first term should be one that defines the *condition* of the soil in the field, in this case, "dense." There are five acceptable terms: very loose, loose, medium-dense, dense, very dense, corresponding to either blow counts of the Standard Penetration Test (obtained during the drilling and sampling operation), or to the relative density percentage, as determined in the laboratory, or

perhaps merely based on the judgment of the evaluator (assuming that neither of the former numbers have been obtained, as might be the case for an inspection of a test pit sample).

If the soil is essentially cohesive (rather than essentially granular, as in the case above), the first term should be the consistency rating, based on either a qualitative judgment or the unconfined compressive strength, if known. (The latter is a laboratory compression test.) There are six acceptable terms: very soft, soft, firm, stiff, very stiff, hard.

If the soil is a borrow material, the density or consistency term is omitted, since the soil will be disturbed and hauled to another site for placement and compaction (if a load-bearing fill).

2. The second term is the color. No limitations on usage of terms are imposed, except that they be common terms. (Fuchsia, mauve, and robin's-egg blue are not recommended.) "Collective" terms associated with visual appearance may also be used, such as blotchy, streaked, and mottled. Color and other visual features may be very helpful to the classification. Munsell charts are often used by scientists in classifying soils. Color and other visual aspects of the soil often provide clues as to the geologic origin, deposition, transport, weathering, and the overall geologic history of the soil.

3. The underlined c̲ may be used to indicate predominance of gradations; in this case, the evaluator judged that coarse sand was more in evidence than the other sand gradations. Thus, c,m, and f sands are present, but most of the sand is coarse.

4. Some(-) merely means that the classifier (obviously a show-off) judged that the percentage of fine gravel is closer to 20 than to 35.

A rule that I have added to the Burmister system (I would not presume to change it) is the suggested use of the semicolon as a flag that the formal description is complete. Anything following the semicolon is free form—write whatever you want and judge might be helpful to the reader. Before the semicolon, you must follow the rules assiduously.

I recall that during the last several years of my teaching career, I would often say to my students that what comes after the semicolon in the not-too-distant future may very well be more important than what comes before, because of the increasing emphasis that was then developing regarding environmental matters, particularly in developed, industrial areas such as northeastern New Jersey. I guess my observation was more prescient than I thought, as there is now, believe it or not, a *Journal of Soil Contamination* (Dragun, 1992).

What one chooses to write after the semicolon is, of course, a matter of judgment, but recognize that the sample that is described before the semicolon in accordance with the rules of the Burmister system, refers only to a very small sample of soil, either extracted from the split spoon sampler (for an essentially granular soil) or from a Shelby tube (for essentially cohesive, plastic soils), or taken from the wall of a test pit (a "chunk" sample). Accordingly, a lot of very useful—even critical—information may be observed about the surrounding site. Thus, in the example description, the "rounded gravel," and the added observation, "occasional to frequent rounded boulders," probably connotes an alluvial soil (deposited by flowing water—hence the rounded shape). The presence of boulders indicates the likelihood of it having been a "high-energy" flow—a flood, perhaps, or a glacial outwash—that was responsible for the transportation of such large particles.

Finally, I would recommend the use of qualifying adjectives like "faint" (or "distinct"), and the use of question marks to indicate to the reader that the field person had some doubts about the written observation.

The grain size distribution curve for the example is plotted in Figure 9.3. Since the description was by visual estimation, rather than by a sieve test, the plot is an envelope rather than a single line. (If the classification was well done, a sieve test would plot somewhere between these limits.) Point 1 represents the maximum size particle present in the sample, f gravel. Points 2 and 3 represent that 65–80% of the sample is finer than gravel (since "some" means 20–35%). Points 4 and 5 represents the 0–10% (trace) silt. Points 2 and 4, and 3 and 5 are connected by recognizing that the c gravel predominates. Note that the curve will be steepest in the range where this is the case. For a helpful example, note that a horizontal line through a grain size fraction means that *none* of that particular size fraction is present. This occurs in what Burmister called "skip-graded soils."

Three indicators relating to granular soil were adopted by Burmister: the D_{10} size, also called Hazen's effective size, the range of sizes C_R, and the curve shape.

The effective size was adopted from the work of Allen Hazen, a prominent sanitary engineer who studied the permeability of soils as related to sand filters in water treatment operations (see p. 49). In Figure 9.3, D_{10} is about 0.035 mm. Note from p. 49 that an empirical formula allows one to calculate an approximate coefficient of permeability. This feature alone may make the Burmister system very attractive for use in specifying acceptable texture for select fills for mound construction.

The range of sizes is simply a number approximating the number of grain size fractions over which the soil extends, expressed to the nearest 0.1. Burmister explains techniques whereby a straight line representing the curve is constructed, from which the C_R is obtained. For the curve of Figure 9.3, C_R is about 5.0.

Burmister examined many grain size distribution curves (probably thousands) and detected five characteristic shapes: S, E, D, L, and C, with some combinations and variations. S is characteristic of most sands, C of most gravels, L of silts, and E and D of skip-graded soils. For illustrative purposes, a couple of these curve types are plotted in Figure 9.3.

APPLICATION TO MOUND DESIGN I believe that the Burmister system can be effectively applied to the selection and specifications for select fills for mound systems. Because of what I would consider an "unacceptable exactness" (to be polite) of systems currently in use, I believe that so-called "select fills" that currently get into mound systems can often be of almost any texture, largely because of the absence of any meaningful specification criteria, which is of course linked inextricably to the lack of an adequate classification system. The Burmister system, I think, fills this need. One way to accomplish this is to specify a range of acceptable D_{10} sizes. Another is to specify an envelope of grain size distribution curves into which the proposed select fill must fall. The latter approach is commonly used in designing filter materials for structures such as earth dams and levees (Cedergren, 1990). Thus, it would be sensible to extend this established technology, with appropriate modifications, to the design of mound fills. (As I write, the Great Floods of 1993 are ravaging the midwest, and levees are failing under the surge of these historic events. Thus, it is reasonable to assume that the design, restoration, and maintenance aspects of such earth structures will get a lot of attention in the years ahead. I would suggest that the reader might wish to pay particular attention to Chapter 6 of this book.)

ASPECTS OF CONSTRUCTION Once a fill has been chosen for the mound (hopefully one that has suitable texture based on Burmister classification criteria), there are certain do's and don't's that must be observed during construction to ensure that the finished structure will perform satisfactorily. The Wisconsin paper (Converse, 1975) provides some excellent advice such as ensuring that the soil is dumped at the perimeter of the mound area by the heavy trucks that deliver the soil. The dumped soil should then be spread into place by a crawler tractor rather than a wheeled vehicle. Both of these precautions are to avoid rutting and compaction of the soil by heavy, concentrated wheel loads.

One construction goof that I learned about at the Cook College seminar is the inadvertent creation of "DriSSCs"—dribble stone solution channels. The acronym is mine, but the term "dribble stone" was coined by a speaker at the seminar, Edward Mott, who was listed as a Wastewater Specialist, but who introduced himself as a contractor. He described a case where the builder of a mound had allowed stone that was supposed to be placed on top of a mound to dribble down the slope of the mound, where the stone accumulated at the base. Instead of removing the stone, it was buried by the select fill material. This, of course, created what in geologic parlance is called a solution channel. The stone that accumulated would, in nature, be called "detritus" or "talus." The solution channel thus formed would behave much like a French drain. And obviously the mound would not behave anywhere near the manner in which it was intended. In Section 9.3.4, I shall share some additional comments gleaned from the very useful and helpful presentation of Ed Mott.

The final element of the mound system, top to bottom, Figure 9.1c, is the natural soil. This surface should be stripped and grubbed (the removal of vegetation, including stumps and roots). As was recommended for trench bottoms, the surface should be raked and cleaned to remove the effects of any and all traffic compaction.

SUMMARY AND RECOMMENDATIONS What follows pertains to mound systems and the problem of select fills for mounds, but much of what I say will have a broader connotation, namely to generate the action that will be necessary to get all engineers to adopt and use the Burmister system of soil classification.

In order of importance, I would say that the following constitute the essential features of this section. First is the overall adoption of the Burmister system. Within that, the use of the D_{10} size, and/or the established technology of using an "acceptance envelope" of grain size distribution curves for choosing select fills for mounds. For evaluating and classifying fines, the dilatancy test and the PI (and field tests to estimate same) are recommended. Incidentally, none of the above was mentioned at the Cook College seminar. It is obvious, at least to me, that the input of soils engineers is badly needed, and I hope this section will serve the beginnings of that effort.

The first and very important step toward the possible acceptance of the Burmister system is to see that the description of the system gets the wide attention that it deserves. But we have a problem. Burmister's *Soil Mechanics*, which was published by Columbia University in 1955, is out of print. However, Professor Burmister published complete descriptions of his system in *ASTM* (1953, 1958), which were even more detailed than was the treatment in his Columbia volume. These, too, are out of print, but should be available in most of the libraries of engineering schools and other technical libraries.

In the mid 1950s, yours truly took his first soil mechanics course at Newark College of Engineering, given by Leonard Shapiro, who had written an in-house publication called "Soil Mechanics." That publication is also out of print. However, in subsequent years, I developed my own set of notes into what eventually became a manual called "Soil Mechanics Laboratory: Procedures and Write-ups." It is about 90 pages, and contains as an appendix a short manual that was used by the employees of a consulting soils engineering firm for which I worked, Woodward-Clyde-Sherard. I would be willing to make available all or any part of this manual to any interested reader, with arrangements to be completed at a later date, such arrangements being P/H/G. (That's postage, handling, and greens fees.) The pertinent sections of the manual relating to this work are "Procedures for Soil Identification and Classification," and the W-C-S manual, "Notes on Soil Sampling and Field Identification." (The "Procedures . . ." includes a two-week segment which covers, sequentially, granular soils and cohesive soils. The latter, I understand, came to be known among students as "Mudpies II.")

I also offer my services as a consulting lecturer on the techniques herein described. All the careful writing in the world will not compare to the value of demonstrating the techniques, and having the audience participate in hands-on activity prior to follow-up field experience needed to develop the skills. Interested parties may contact me at 85 Newark Avenue, Bloomfield, NJ 07003. Phone (201) 743-6210.

Regarding the field experience mentioned above, I have been reflecting on my own early experiences as a young practitioner with Woodward-Clyde-

Sherard, starting in the summer of 1961. One of the unique and (in retrospect) very valuable aspects of the overall experience was related to the fact that W-C-S maintained their own drilling and sampling operations and their own laboratory. I would venture a guess that no such situation exists anymore, so the experiences that I shared with my colleagues at the time will never be repeated by others. Too bad.

One of the memorable (read harrowing!) experiences was the first time I was assigned to be the engineer on a drilling and sampling operation. I was issued the manual mentioned above, and was told "Do it!" While I had been teaching for three years, I had not yet become a soils specialist by any means, and had not committed the rules of the Burmister system to memory. Picture the situation! The drilling crew is evaluated on the number of borings (feet drilled), and here they are stuck with a rookie engineer who proceeds to leaf wildly through the manual as the split-spoon samples come out of the ground. Somehow or other, I muddled through, and even stayed friends with the guys on the drilling crew. I have the soil-stained, original copy of the manual to prove it. Needless to say, I was one tired dude at the end of that first day! The experience, however harrowing at the time, was invaluable.

Another equally valuable experience was my association with Bill Mills, the man in charge of the laboratory. Bill was, in my opinion, something of a genius when it came to laboratory testing. He had an uncanny ability and interest in designing tests to fit special one-of-a-kind field situations. Anytime you'd go into the lab, he'd have some contraption set up to test a soil in some special way. (As soils engineers are aware, there are many highly standardized tests for determining soils properties, and Bill knew all of them, but he loved his one-of-a-kind contraptions.) As I recall, Bill never graduated high school, yet he went on to become a partner in a major soils engineering firm.

Bill was the guy who showed me how to do the dilatancy test. It was, I think, the first question I asked him. And I think he was pleased that I asked. He took me under his wing thereafter, so to speak. My philosophy, for what it's worth to you young engineers out there, is to get help from the best source and, above all, don't be embarrassed to ask "dumb" questions. The only dumb question is the one you were afraid to ask. Bill Mills passed away some years ago at a fairly young age. Too bad. He was quite a guy, and he would have enjoyed this story.

My last recommendation in this section is that some bright young engineer write a text book on introductory soil mechanics, hopefully one that will be adopted at many engineering schools across the country. If that occurs, the rest will take care of itself. In that book, of course, would be a strong presentation of the Burmister soil classification system, and a list of reasons as to why it should be adopted as *the* soil classification system. I would further hope that whoever might produce this text would be practice oriented, rather than the more typical academically oriented textbook author.

In the latter part of my academic career, I taught a senior technical elective course which was case history oriented. Such a course gets students more ready for the real world of engineering, and also allows for the further development of communications skills, because, rather than assigning homework problems and giving tests with textbook problems, the assigned work was the preparation of engineering communications, including letters and reports. I believe a textbook prepared for a course like this (called by some "guided design"), with a lot of case history material from the author's professional experience, would be an excellent forum in which to present the Burmister system.

Columbia University still teaches the Burmister system and it is used by many engineering and consulting firms. Thus, there is considerable evidence that the system is highly favored in major areas of academe and professional practice. All that is needed, perhaps, is a big push to get it accepted universally. A top-notch undergraduate textbook on soil mechanics that espouses the system would go a long way toward fulfilling the reality.

Professor Burmister was a faculty member at Columbia for 34 years, and a pioneer in the field of soil mechanics. He set up the first soil mechanics laboratory in the United States in 1933. During his career, he investigated earthworks and foundations for over 400 projects, including such famous structures as the Verrazano Bridge. He published more than 35 research papers. Professor Burmister died in 1981. His students remember him as a kind and patient man. He was, as one of his colleagues said, "way ahead of his time."

9.3.3 Problem Sites

It is beyond the scope of this book to describe in detail all options that are available for coping successfully (one would hope!) with what I lump into the overall category of problem sites. As has been inferred, these would be those with a relatively shallow depth to some limiting zone (in Pennsylvania, less than 20 in.). I shall, however, list many alternatives that are available, provide brief commentary on each, and list references for those who wish to learn more. The one thing that most of the alternatives have in common is that they are more expensive, at least initially, than standard or mound systems, and this is to be expected. Before proceeding, I might note that Pennsylvania designates anything other than standard or mound systems to be experimental, and each must be evaluated on a case-by-case basis. Whether that is the practice in other states, I can't say, so the first step by any reader who may have a problem site is to investigate that aspect. Second, it would behoove anyone other than an environmental engineer . . . we used to call them sanitary engineers . . . to seek the help of a professional in evaluating possible alternatives. This, of course, is one of the expenses that adds to the increased cost of the process. Not a small part of the problem will be the shepherding of the proposed solution through the myriad of regulations (*and* regulators!) that are sure to be lying in wait along the way.

I shall attempt to present the alternatives in the perceived order of cost and difficulty. Understand that these are just my perceptions, and no guarantees of absolute accuracy is to be inferred.

1. *Give it up.* This advice may seem curt or flippant, but may turn out to be the best alternative, and should be regularly considered and compared to the cost of specific other options. As good poker players know, sometimes it's better to get out early, if you don't have a good hand. Stated another way, don't throw good money after bad. Advice from a professional should be sought, but even here, there can be no guarantees of success.

2. *The Infiltrator system.* Although quite new, the system seems to have considerable promise, for both effectiveness and cost saving. Indeed, the system may even be less costly than conventional standard and mound gravel systems, particularly if the claims for reduced size of field are proven to be even approximately correct.

During the latter stages of research for this chapter, I have been in communication with the co-inventor of the Infiltrator system, Randy May. He has sent me publications that include a great deal of documentation that illustrates the effectiveness of the system as compared to gravel systems, much of the documentation by independent investigators.

In a conversation that I had with Mr. May, he indicated that there was some evidence that the use of filter fabric (geotextiles) with the Infiltrator units has not been completely favorable . . . that some failures had occurred. I'm not sure I understand what the reasons were thought to be for these failures, but I suggested to him that he contact Robert M. Koerner, Director of the Geosynthetics Research Institute (GRI), Drexel University, Philadelphia, PA. Bob Koerner is also the co-author of the definitive book on the use of geotextiles: *Construction and Geotechnical Engineering Using Geotechnical Fabrics* (Koerner and Welsh, 1980).

I do not know whether the GRI has done any research into the use of fabrics specifically dealing with septic system applications, but I have a suggestion that I am virtually sure will be a new area of investigation; *Tabasco sauce! Now, after the guffaws have subsided, let me explain. On or about 4 May 1993, I heard an "Osgood Report" on CBS radio that featured an inventor who got an idea for an invention at a cocktail party (my kind of inventor!) where he was served deviled egg with a generous dose of Tabasco sauce.* In the interview, the inventor said, "it about fried my eyeballs," or words to that effect. But, as Charles Osgood observed, it did not much hurt his brain, since he got the idea to develop a paint for the hulls of boats containing ingredients similar to those in Tabasco sauce, reasoning that barnacles, being living organisms, would not be happy about snuggling up to hulls treated with the special paint. According to the report, the extract of jalepeno peppers was one of the ingredients. Field trials in waters off Florida have been successful, and have led to a patented product called Barnacle Ban.

It occurred to me that, since bacteria are also living organisms, and thus probably wouldn't like jalapeno peppers any more than barnacles do, and *if* the failure problems of filter fabrics in Infiltrator septic systems are caused by some form of biologic (algae?) growth, fabric sprayed with Barnacle Ban might be a viable solution.

The address of the Infiltrator company is 123 Elm Street, Suite 12, Old Saybrook, CT 06475. Phone (800) 221-4436.

3. *The Four-Year Fill Option.* This option is available under current regulations in Pennsylvania. Whether a similar option exists in other states is not known. The procedure involves the construction of a fill at the proposed location of the eventual system. As far as I am aware, there are no published specifications for the fill. I would suggest the following: (a) Choose the best location on the site with respect to existing natural soil and topographic considerations. (The guidance of a professional will be most advisable for this and other steps.) (b) Strip, grub, rake, and clean the site location, being careful to avoid undue trafficking on the site. (c) Place a fill of a texture that has a marginal permeability to a height of 30 in. (See earlier section regarding selection of fill texture.) Use machinery and methods that will minimize compaction. (d) Add sufficient topsoil to support grass growth. Seed the topsoil. (e) Wait four years! (f) Proceed with the evaluation of the prepared fill in the usual manner, i.e., with test probes and percolation tests.

This option, quite frankly, puzzles me to some extent. It appears to be simply a matter of providing by construction that which nature failed to provide. The puzzling part is the four-year wait, a requirement that seems to have no rational technological basis for its imposition other than some minor and inconsequential settling. I have asked a number of regulators for the reasons, and have never received a good reply, although one regulator conceded that the four-year wait may be nothing more than an attempt to discourage people from exercising the option.

4. *Rent (or buy) a leach field from a contiguous property owner who owns a lot that has been approved for septic.*

5. *Shared solutions.* Locate a site within reasonable distance from the problem site where soil conditions are suitable for septic. Often, there will be others near you who own problem sites. A combined system may be possible, wherein the effluent is pumped from the homes to the "remote" site for discharge into the septic system constructed there. In effect, this constitutes a mini-sewerage system, one that will obviously have to be designed by a professional. The cost would probably be substantial, but perhaps within reason on a shared basis. This procedure has been done successfully in the Pocono region.

6. *Aerobic systems.* The standard and mound systems described are known as anaerobic systems, in that all of the biological action in the septic tank takes place in the absence of oxygen. If the effluent is subjected to oxygenation in some manner, the system is said to be *aerobic.* One manufacturer of such a system asserts that their system, the Rotordisk, has an efficiency rate of 90–95% in removing pollutants, leaving only 10% of the job to be done in the drain field,

as opposed to the 60% that the field of an anerobic septic tank must do. With purer effluent from aerobic tank treatment, the drain field can be much smaller, as much as two-thirds smaller, according to another manufacturer (Cromaglass). Thus, aerobic systems may be suitable not only for sites that have problems with respect to subsurface conditions, but also where space limitations are severe.

The National Sanitation Foundation (NSF), Ann Arbor, MI, publishes lists of products for individual aerobic treatment plants of 1500 GPD (gallons per day) or less that have been approved under the provisions of NSF Standard No. 40. The May 1986 list was, in turn, issued by the Department of Environmental Resources of Pennsylvania as a supplement to their Chapter 73, Standards for Sewage Disposal Facilities. So it appears that many states would accept the ratings and judgments of NSF. Company names and addresses are also included. The Rotordisk and Cromaglass systems appear on the 1986 list of NSF, and I have corresponded with each of these companies.

Rotordisk is described by the manufacturer as a mechanically simple device, consisting of a shaft (with disks) driven by a small motor. The disk assembly rotates half-submerged at a slow speed (3 rpm). During the part of the rotation in the air, the biological action on the pollutants is enhanced (as contrasted to that of anaerobic action in standard tanks). Installation and maintenance are simple and easy, and the manufacturer claims the units are quiet and odorless. Also, cold weather "is no problem." The unit is encased in a manifold, and protrudes less than 1 ft above ground. Its plan dimensions are about 6 by 7 ft. These data are for an S12 model in a fiberglass tank, a size suitable for a single-family residence. The cost quoted was $5637 (U.S. dollars, 1988). The present, installed cost would, of course, be significantly higher. Address: CMS Rotordisk Inc., 140 Snow Boulevard, Unit 3, Concord, Ontario, Canada L4K 4C1. Phone (416) 625-8916.

Cromaglass features a unit that pumps effluent through a series of treatment chambers, including an aeration chamber. Their claims are similar to those of Rotordisk with regard to effluent output purity and low noise and odor. I did not obtain any cost figures. Address: Cromaglass, Williamsport, PA 17701. Phone (717) 326-3396.

A complete listing and description of all of these systems is beyond the scope of this book (and, indeed, beyond the expertise of the author). Readers are encouraged to work with a professional in the possible selection of a suitable aerobic system.

A caveat: One source that I discovered in my research asserts that, in general, aerobic systems "are not effective for weekend homes [because] they must be worked at all times in order to be effective" (*The Pocono Record*, 19 August 1976, p. 3).

I wrote to the manufacturers to request their response to this caveat. Tom Smith of CMS Rotordisk (he is the 'S' in CMS) responded with a convincing array of data and studies that were specifically undertaken to investigate the effects of intermittent use, and the results were affirmative. In fact, it is claimed that "a system that is underload will produce a better quality effluent than a system that is at design loads" (Smith, 1993). Indeed, data from an independent study by the National Sanitation Foundation (NSF), a noncommercial agency, list removal efficiencies of 93 and 96% reductions of 5-day biochemical oxygen demand, and suspended solids, respectively, in the treated effluent. Apparently on the basis of these kinds of results, the claim is made that "the high quality effluent can be used for . . . *groundwater recharge*," (emphasis mine), and that "lots that were marginal on the basis of individual septic tank systems can be developed." (Rotordisk, 1993).

7. *Thetford Cycle-Let System.* This wastewater and recycling system has its principal applications in commercial buildings and developments, shopping centers, and schools. By virtue of recycling much of the water, discharges can be reduced by as much as 95%. Because of the nature of the wastewater in residential applications (only about 35% from toilets and urinals, as compared to 90% for nonresidential buildings), the system is not cost effective in single-residence applications. The manufacturer indicates that the system "is suitable for condominium developments, housing projects, hotels, etc." The Thetford Corporation address is P.O. Box 1285, Ann Arbor, MI 48106. Phone (800) 521-3032.

9.3.4 Toward Improved Design and Construction

I would like to close this section with some brief but, I think, important comments about some simple steps that can be taken to improve design and construction decisions and procedures. My first suggestion is that the designers talk with the contractors who build the systems, and the second is to encourage "postmortems" on failed systems to find out, if possible, what "killed" them.

As an example of the potential value of the first suggestion, I relate a conversation I had with a contractor of long experience in the Poconos. (He is the same one who experimented successfully with a trench system, as earlier described.) He observed that the holes in the perforated PVC pipes of the distribution system (Figure 9.1*c*) were 3/16 in. diameter, and "should be twice that." He also commented that the holes had burrs, and thus tended to snag a variety of solids that had found their way to these outlets, being caught by the burrs. These solids, I would guess, might include toilet paper fragments, nail clippings, and cigarette filters, along with all manner of other objects that never should have been thrown in the toilets and sinks, but commonly are, because of the lack of advice given to owners about simple do's and don't's about septic system maintenance. It occurred to me that these problems could be easily mitigated at little cost, by a simple design change (hole size), by materials specifications (a simple instrument could remove the burrs), and by greater attention to simple maintenance education directed toward owners.

The idea of "postmortems" follows from the foregoing. I believe a lot of useful information could be gleaned by such a regular practice. I would expect that most contractors have a fair amount of experience repairing or replacing deficient or failed systems. Inspections of the excavated soils and system components could yield a lot of useful indications of what caused the problem, and steps could then be taken to avoid the same problems in future design and construction.

As mentioned in a previous section, a constractor/wastewaster specialist, Ed Mott, was one of the speakers at a recent seminar I attended on Septic Systems at Cook College, Rutgers University. His first comment was "I'm a contractor. I think what we need to do is communicate and get along." I thought, this is my kind of guy!. He then proceeded to share his thoughts on what makes systems fail. Here is a summary, as compiled through my recollections and notes.

He likened a septic system to you and me. (While that was not intended to be funny, when's the last time someone called you a septic system?) He suggested that, just like a person, a septic sysem needs air, water, and proper food to remain healthy. However, too much water and you drown, or at least get sick, because you've been deprived of air for too long a period. Since the processes that go on in both a septic tank and the drain field involve living organisms (bacteria), the analogy is not really far-fetched or inaccurate. You need food, the bacteria need food. The bacteria's food is the waste materials that are contained within the effluent. If you ingest poison, you get sick, and maybe even die. If you poison the bacteria by disposing of "poisons" in the effluent, your septic system will get sick or die. Make sense? I thought so.

Mr. Mott also gave strong endorsement to the use of pressure distribution designs, pointing out that all drain field surfaces, no matter how carefully constructed, must have a low spot. After all, I jotted in my notes, we're not talking here about Willy Hoppe's billiard table. (For you younger folks out there, Willy Mosconi? Steve Miszcerak? For those of you completely lacking in the cultural niceties, these are all champion billiard players of different eras.) Thus, if gravity flow design is used, the effluent will always find its way to the same low spot, and be concentrated over a tiny area. Eventually, that area will become clogged by "bad things" in the effluent, and the effluent will start to seep into the perimeter area of the initial clogged area. In this way, the clogged area will grow, and in time, the whole system will have failed. Thus, this seems to be a good case for the efficacy of "forcing" the flow over the entire drain field, as is the design intent of pressure distribution systems. Of course, no system will survive long if badly abused by the poisons that are too often imposed upon the systems through poor habits of the owners. (Much more on this in Section 9.5, Maintenance.)

In correspondence with Ed Mott, he supplied some observations that strongly support my contention in Section 9.2 that a better system of soil classification is needed for septic system design. He pointed out that "bank run in New Jersey is frequently used under the stone of sewage disposal fields." Figure 9.4 is a photograph of a mound, courtesy of Ed Mott. It shows the "dribble stone" on the slope of the mound and illustrates the tendency of the

Figure 9.4. *Dribble stone in mounds.*

material to accumulate at the base of the slope, where it might be buried. The "bank run" mentioned by Mr. Mott is shown in the photo as the material immediately beneath the surface layer of coarse stone. Mott asserts further that the material

> consists of everything from clays to large rocks. When dumped . . . the larger particles tend to flow to the outside of the pile . . . being placed. This separation of larger particles causes voids or channels to form vertically [unintended internal "chimney drains"]. When sewage is applied in large enough quantities to cause saturated flow, these channels conduct the effluent deeper into the earth further away from the oxygen source nearer the surface and frequently into groundwater without having received the necessary and beneficial aeration treatment.

Mott mentions the "uniformity coefficient." This is an index property that is part of the Unified Soil Classification System. It is defined as the ratio of the D_{60} size to the D_{10} size. A uniformity coefficient of 1.0 would connote a soil containing all the same sizes. Obviously, the larger the uniformity coefficient, the wider is the distribution of sizes. This coefficient would represent still another means of specifying acceptable select fill material for mounds. Note that this index is analogous to Burmister's range of sizes.

Mott emphasizes the importance of avoiding saturated flow, which in turn confirms the need for water conservation. As noted, this important topic is covered more fully in Section 9.5, Maintenance.

At the Cook College seminar there were a number of comments by professionals about "bank run" material and its implied "fuzzy" connotation. Some even mentioned that some earthwork "entrepreneurs" were advertising that their soil was from a "certified pit." None of the folks from the state had ever heard of any such certification. So, beware. Some earthwork contractors that I know of have large-scale screening capabilities. This seems to me to be the way to go, particularly after some meaningful criteria have been adopted to define acceptable texture for mound fills, some of which have been suggested here, based on the Burmister system of soil classification. Undoubtedly, such soils will cost more, but the money will be well spent, in my opinion, for it will further ensure a long-lasting, trouble-free septic system, and one that will not pollute groundwater.

9.4 RENOVATION/REPAIR OF MARGINALLY PERFORMING SYSTEMS

I suspect that there will be a lot of avid readers of this section, because of a companion suspicion that there are a lot of systems in trouble out there in septic land! Unfortunately, the task of renovating or repairing marginally performing systems is a daunting one, since almost none of these systems were designed with easy accessibility in mind; like doctors, I guess, septic designers and builders bury their mistakes. I have long believed that the criterion of accessibility should be an important one. Fortunately, for future construction, there now exists a system that is much more accessible than all others heretofore: the Infiltrator System. Indeed, it occurs to me that the system could be further improved in this regard by using custom-molded foam plastic blocks as "backfill" above the Infiltrator units, in which case geotechnical filter fabric could be attached over the slotted sides of the units. The top could then be lined with some suitable outdoor carpet, or perhaps with something like Astroturf. Such a modification would produce a system that would involve no excavation, should a need for inspection and repair arise.

As for existing systems, the term "marginally performing" was chosen to reflect the fact that systems usually do not fail suddenly: rather they gradually exhibit signs of partial clogging, for example, by an inability to do several loads of laundry in succession. When such signs develop, often after years of poor or nonexistent maintenance, you may be in need of renovation, repair, and/or (perhaps a better term) septic system *rejuvenation.* In such an event, I would recommend consideration of the idea of straddling the existing system with an Infiltrator system, assuming that space permits. This would create what is called "alternating systems." As the name implies, you would then have two systems that can be used alternately and independently by virtue of a plumbing system that would allow one system to be used for a period while the other rests.

Kaplan (1991, p.265) indicates that "an old, failed [damaged] leachfield may be used again *after* it rests for three to five years." After the 3- to 5-year rejuvenation period, Kaplan recommends that the systems be used on an alternate-year basis. Obviously, other systems could be used as a new alternate, but may be more expensive and less accessible.

If space does not permit the alternate system, it is likely that a new one would have to be built in its place, requiring the expensive removal of the old system. In such a case, it would seem logical that excavation be extended to a level beneath the clogged old field, thereby exposing fresh natural soil. This time around, *pay attention to maintenance!!*

9.5 MAINTENANCE

The importance of maintenance has been repeatedly stressed in earlier parts of this chapter, and is reiterated here for even greater emphasis. I believe that if only a tiny fraction of the money and attention given to micro-regulation was instead directed toward the education of owners of septic systems about how to take proper care of the systems, the net result would be far less pollution of the environment, especially groundwater. I believe further that some teeth have to be put into the education process, by passing legislation that *requires* that septic system owners be apprised, in writing, of good maintenance practices, almost all of which are simple and preventative in nature. This could easily be done through the use of inserts in other mailings so that the cost of mailing would not be a factor. Lastly, as described in a previous section, I am convinced that almost all owners want good information badly, and in their own self-interest will respond positively to intelligent suggestions regarding preventative maintenance.

A step in this direction was taken by Kaplan (1991, p. 265), when he prepared and made available for distribution a leaflet on good maintenance. These are distributed by private business people (septic pumpers, realtors, etc.). Kaplan starts his leaflet as follows: "This notification could save you anywhere from about $600 to the actual value of your home." Now, if that doesn't get your attention, consider the following eyeopener. By remarkable coincidence, this startling article appeared in the local paper on the same day I read Kaplan's leaflet (*Pocono Record*, 1993).

> *Warren, PA (AP)* DER [Department of Environmental Resources] is asking townships in the state to adopt an on-lot management ordinance. Under the ordinance, townships would hire a sewage inspector. DER claims those sewage inspectors would be empowered to go onto private property, demand information, make sewer system tests, go inside buildings, dig up yards, and even order the occupant to move out—all without the landowner's consent, or warrants." A solicitor for several townships asserts that this is "unreasonably intrusive and unconstitutional," and that "DER is asking the township sewage inspectors to assume powers that even state and federal police do not hold."

I say this is regulators running amok. Legislators, please take note of what you have wrought! With that, I trust that readers will now pay attention!

I recommend Kaplan's leaflet, but I am going to get considerably more specific about the do's and don't's of septic system maintenance.

9.5.1 Septic Tank "No-No's"

I suggest three categories of items that should never be allowed to enter the septic tank through any of the five ways that they commonly do: toilets, tubs, sinks, washing machines, and automatic dishwashers. The categories are floats, oils, and suspended solids, and their first letters form the acronym, FOSS, a memory device that I think will be helpful to the reader in evaluating objects that may be detrimental to the "health" of their septic system.

Obviously, anything that floats has a potential for entering the drain field, and possibly clogging either the soil or the perforated holes of the distribution pipes. Among items that I can think of that should be avoided are cigarettes (especially the filters), matches, dental floss, nail clippings, and hair. The latter can easily be avoided by using an inexpensive strainer that fits over the tub drain hole, available in most hardware stores.

Oils are especially to be avoided, or at least minimized. Oils, of course, float, but they are so potentially damaging, that they deserve a separate category listing. One easy and effective way to reduce the amount of oils entering the system is to get into the habit of wiping all oily objects or utensils with absorbent towels, and discarding them in the trash can. (Landfills can more easily absorb these than can your tiny drain field!) Do not pour even small amounts of oil (for example, from a sardine can) into the sink (something I once did routinely, I'm embarrassed to admit). Instead, place a paper towel in a suitable dish, and pour the oil into the towel. Then discard the towel and the can in the trash can. Wipe thoroughly all dishes, pots and (especially) frypans that contain obvious quantities of oily materials. Wipe the percolator basket that holds coffee grounds; a surprising amount of oil from the grounds lurks here, and every little bit hurts.

Suspended solids, as the name implies, are so fine that they remain in suspension for long periods of time after they enter the septic tank. When periods of high water usage occur, as is sure to happen to some degree, the water containing the suspended solids will be discharged into the drain field, where the solids will eventually settle and contribute to the clogging of the field. By far, the greatest source of suspended solids from the home is the sink garbage disposal. This has been convincingly documented by Perkins (1989). In a table entitled, "Wastewater Characteristics" (p. 194), the combined amount of suspended solids for toilet, bathing, kitchen sink, and clothes washer, was 1300 mg/L. For the garbage disposal, it was 3500 mg/L, almost *three times as much* as that from all other sources combined!! The sensible recourse is clear: *Do not use a garbage disposal.*

HARSH CHEMICALS The term "harsh" is used here to reflect the fact that it is impossible to keep *all* chemicals out of the effluent. Consider, for example, the following list: stearic acid, triethanolamine, isobutane, laureth-23, sodium lauryl sulfate, and propane. From some exotic top secret military chemical weapons research project? Nope, it comes from the label on my shaving cream (Barbasol).

The problem confronting the homeowner is to decide which chemicals are harsh, and therefore to be avoided. Easy question, not so easy answers. This is so because there are no definitive or completely reliable answers to what products are "septic safe." As far as I know, there are no federal regulations covering this area, at least not yet, so product manufacturers can ignore this, or worse, make claims that are not proven and may not be accurate.

I think that, in time, more activity will develop toward getting lists of septic-safe products. Perhaps there already are such lists that have not yet been widely published. In the meantime, septic system owners will have to use common sense, intuition, and perhaps a little reading of their own (labels included) to make intelligent decisions. Following are some comments about various chemicals and product types that are best avoided, and also a suggestion that will reduce the amounts of chemicals that do enter the septic tank.

Most authorities suggest the avoidance of bleaches and phosphates. A lot of cleaning products are now available that are phosphate free. Phosphates are no-no's because they promote algae growth.

Avoid toilet papers that are other than white. Scented and colored or patterned paper contain both dyes and scent chemicals.

A suggestion: Whatever products you do decide to use, use half as much. Years ago, a friendly garden store owner told me that I should set my fertilizer dispenser at half the recommended setting, that the manufacturers were stipulating much higher than needed settings simply to sell more fertilizer! Since then, I have tried the "one-half rule" on a lot of other products, with perfectly acceptable results, in general. You will not only reduce the amounts of chemicals entering your system, but you'll save a few bucks to boot!

Lastly, people who dump things like crankcase oil into the system are a few croutons short of a salad. Here is a list of some additional items that should not be allowed to enter the septic system: coffee grounds, cat box litter, facial tissues, paper towels, and bulky wastes that you would judge not to be biodegradable, such as plastic and glass materials, rubber, and nylon. Also, sanitary napkins, tampons, and disposable diapers. Hazardous chemicals would include paints, varnishes, thinners, photographic solutions, poisons, pesticides, and herbicides (Perkins, 1989; Sponenberg et al., 1985).

9.5.2 Water Conservation

Each time the effluent level in the septic tank reaches the elevation of the outflow port, effluent will flow to the drain field. If the tank is kept to this level by excessive use of water within the home, the drain field will never get a chance to "rest." During periods of heavy rain or snow melt, additional water will enter the field and the tank. As I write, for example, the northeast is recovering from the "Blizzard of 93." It is late March and the Poconos are covered with the melting remains of three feet of snow, also, of course, completely blanketing the drain field and the septic tank. This is *not* a time to be washing all of the sheets and towels, or luxuriating in two-foot deep hot baths. Instead, it is time for all manner of water conservation measures, like Murphy baths and Navy showers (explanations to follow).

Many steps can be taken to reduce the amount of water usage, most of which are easy and either cost-free or inexpensive. Indeed, many will save money in the long run, particularly if their combined effect contributes to the indefinite trouble-free functioning of the septic system.

Some of the suggestions are extreme, but I think it best to consider the worst-case scenario, that of a marginally functioning septic system, caused most likely by years of abuse and a period of heavy rain or snow melt on the field and tank. In a more normal setting, the reader can decide which steps are warranted.

Bathroom water usage is the largest contributor toward total household water use, accounting for 70% of the total, with 39% for toilet flushing and 31% for bathing (Baker et al., 1975). Obviously, then, these are the areas where the largest reductions should be sought. The four elements are the toilet, the tub, the shower, and the sink.

The simplest and most inexpensive way to achieve an automatic, per flush savings is to place weighted plastic bottles inside the toilet tank in such a manner that they will not interfere with the mechanisms. Sand or gravel in the bottles will ensure stability. Do not use bricks, as these will slake and the residue may damage the mechanisms. This step will save a gallon or more of the typical 5–7 gallons of a standard tank.

More expensive steps include the installation of special equipment designed to reduce water usage by substantial amounts, indeed to zero with some systems. Toilet models are currently available that use as little as 1.5 gallons per flush. Another patented device, called Future Flush, features a dual handle and flapper valve that can be installed in existing tanks that allows for a choice of low flush or full flush, and claims to save up to 70% of the water used for flushing. Address: Con-Tech Industries, 104 South Mill, Creswell, OR 97426. Phone (800) 446-5765.

An even more dramatic savings is afforded by waterless toilets. One is called a Dry Flush Incinolet, where the waste is incinerated to ash. Models suitable for small homes cost $1449, ready for installation. The manufacturer makes the point that the product can "extend the life of an outmoded septic system." Address: Research Products Blankenship, 2639 Andjon, Dallas, TX 75220. The other is a system called Magic Flush. This system includes an in-ground tank outside the house (separate from the septic tank) in which a fluid (not water) is contained. A purification system allows for the recycling of the fluid, apparently indefinitely, after each flush. The tank has an external above-ground port from which the waste must be pumped periodically, typically by a commercial vendor with a special tank truck for that purpose. A typical system provides for about 6000 uses between servicing. Address: Monogram Sanitation Products, 1001 Monterey Pass Road, Monterey Park, CA 91754. Phone (213) 266-0101.

These waterless toilets, though undoubtedly expensive, provide a way of allowing for continued use of a marginally performing septic system, by virtue of significantly reducing the strain on that system. Also reduced would be the mental strain of wondering whether or not your toilet will flush during periods of wet weather. This strain could be appreciable during periods of entertaining guests, so the option would be attractive in such a worst-case scenario.

An additional step of a mechanical nature that can be taken is to fix all leaks in faucets and in the toilet tank. Most of these will be obvious, but one may not be: the toilet tank, wherein a small leak may not be audible. A couple of drops of food dye in the tank will show up in the bowl if there is such a leak. Leaks can easily account for 10% of your water consumption. (PUC, 1991).

Large quantities of water can also be saved by changing habits, some simple, some extreme. Here is my award winning list! Note that many of them not only reduce the amount of water used, but also diminish the chemicals entering the septic tank.

1. Flush less. It is not necessary to flush the toilet after each use.
2. Take Murphy baths. As promised, a description: Stand in front of a sink *half* full of soapy water, using *half* as much soap as usual (the "one-half rule"). Apply the soapy water to parts of the anatomy that will be easily recognized by most serious conservationists. Rinse with a wash cloth. Dry. Apply favorite deodorant. (You guys out there can use the morning shave water for this purpose, using, of course, *half* the amount of shaving cream. The excess shaving cream, moreover, can be wiped from the hand with paper, and discarded in the waste basket. This is called "advanced conservation." Are you getting the hang of this?)
3. Take Navy showers. Get wet. Shut off shower. Soap up. Rinse. An additional mechanical device that will make this technique easy and save even more water is an inexpensive low-flow shower head equipped with a shut-off button. Thus, the button feature allows one to turn the water back on at the same comfortable temperature setting as initially adjusted.
4. Do not let the water run while shaving, brushing teeth, or for other similar purposes. Many gallons of water can be saved by adopting this simple technique.
5. Keep a jug of water in the refrigerator for drinking, thus avoiding running the water for that purpose. If you own a second home, as is often the case for septic system owners, a particular problem arises when you return to the home after a long absence. At that time, it is a good idea not to drink the water that has been lying in the pipes and holding tanks. Not only is the taste unpleasant, but the water may, in fact, be unhealthy, particularly if the house is old and contains lead pipes. In this event, purchase a small cloth hose, connect it to the sink upon arrival, and discharge the water out a window or door, allowing the water to percolate directly into the ground, rather than into the septic tank. Such hoses are available for a few dollars.

6. Do not use dishwashers for other than a full load. When using clothes washers, set the water amount at a level appropriate to the size of the load. Again, as a reminder, try one-half the amount of soap, and opt for single rinse cycles, if available.
7. When washing dishes by hand, use an "instant recycling" technique. For example, if you have two pots to wash, pour the soapy water from one into the other. If you use the "one-half rule," you will have used one fourth the amount of water and soap, and so forth. Use a pan of clear water for rinsing, rather than running water from the faucet. Wash vegetables in a similar manner.
8. Use a small plastic bowl to catch water while waiting for the water to get hot. Transfer this water to a large basin and save for other general purposes: rinsing, watering plants, etc.
9. If a large amount of clothing accumulates that needs to be washed and you are in a period of extended wet weather or snow melt, consider going to a commercial laundry to do the washings.
10. Finally, start a contest with your children, with suitable rewards for water-saving ideas inside and outside of the house. This will be a certain amount of fun for the children, you may get some genuinely good and novel ideas, and best of all, the children will be educated and sensitized to good conservation habits.

It is perhaps appropriate to end this section with an aphorism by one of my favorite aphorists, ol' Ben Franklin, himself: "You shall know the worth of water when the well runs dry."

9.5.3 Septic Tank Pumping

The frequency with which the tank should be pumped out depends on the tank size, the number of people regularly inhabiting the house, and, in a broad sense, how careful the owners have been with respect to the many aspects of maintenance heretofore described. Table 9.1 lists frequencies in years for year-round, full-time occupancy. (Second home owners should adjust figures as per their part-time patterns.) Perkins (1989, p. 51) describes a method whereby one can measure when pumping is necessary, should someone in the family enjoy such activity. If done once carefully, one could presumably get a pumping frequency that fits actual use.

Commercial pumpers can be found in the yellow pages, usually under Septic Tank Cleaning. The vendor will arrive at your home, remove the manhole, usually a very heavy concrete plug about 2 ft in diameter. He will then insert a large rubber vacuum hose into the tank, and pump the sludge into his truck tank, and then drive merrily away to some secret place the location of which you probably couldn't care less about. These "honey wagons," as they are often generically called, frequently are festooned with more colorful names and

TABLE 9.1 Septic Tank Pumping Frequencies

Tank Size (gal)	Household Size (Number of People)									
	1	2	3	4	5	6	7	8	9	10
500	5.8	2.6	1.5	1.0	0.7	0.4	0.3	0.2	0.1	—
750	9.1	4.2	2.6	1.8	1.3	1.0	0.7	0.6	0.4	0.3
900	11.0	5.2	3.3	2.3	1.7	1.3	1.0	0.8	0.7	0.5
1000	12.4	5.9	3.7	2.6	2.0	1.5	1.2	1.0	0.8	0.7
1250	15.6	7.5	4.8	3.4	2.6	2.0	1.7	1.4	1.2	1.0
1500	18.9	9.1	5.9	4.2	3.3	2.6	2.1	1.8	1.5	1.3
1750	22.1	10.7	6.9	5.0	3.9	3.1	2.6	2.2	1.9	1.6
2000	25.4	12.4	8.0	5.9	4.5	3.7	3.1	2.6	2.2	2.0
2250	28.6	14.0	9.1	6.7	5.2	4.2	3.5	3.0	2.6	2.3
2500	31.9	15.6	10.2	7.5	5.9	4.8	4.0	4.0	3.0	2.6

Note. Numbers cited are years, and assume full-time occupancy (adjust for part-time use).

Source: Faulk, D., "Maintain Sewage System to Avoid Problems," *The Pocono Record*, JStroudsburg, PA, 11 August 1990.

slogans. Two of my favorites are: "We're No. One in the No. Two Industry" and "The Mountain Doo Company."

Written records should be kept of the dates of pumping, along with a sketch showing where the tank is located. When buying or selling a home, this information should be on hand. It would be prudent, also, to have the tank located and the access manhole exposed for the arrival of the tank truck, so as to reduce the fee charged. Currently, the cost is about $150. (The manhole is usually covered by about one foot of soil.)

9.5.4 Additives

A controversial matter regarding septic tank care and maintenance is whether to add anything to the tank or drain field for the purpose of enhancing its performance. An excellent article on this matter appears in the July 1992 issue of *Small Flows* (Wilcox, 1992). A number of people with extensive experience and expertise in this area are quoted, as are manufacturers of the many products, and the consensus is that there is no consensus. Designers generally are opposed to the use of any such additives, declaring them useless. Manufacturers, on the other hand, insist that they work (at least *theirs* works). Complicating the problem greatly is the fact that there are literally hundreds of such products on the market, and there is little standardization for testing or regulations governing manufacturers' claims. So it can all be quite bewildering to the consumer.

According to Wilcox, "Additives fall into two major categories: chemical additives (acids, bases, flocculents) and biological additives (enzymes, bacteria, yeast). Chemical additives are primarily oxidizing agents. Biological additives are designed to enhance biological activity inside a septic tank."

A fairly common opinion among designers and sanitarians is that, of the two types, chemical additives pose the most danger, in that they may go through the tank and the drain field, and enter the groundwater or nearby wells. Some of these chemicals are known to pollute the water and may be serious health hazards, even carcinogenic. Biological additives are thought by many to be ineffective, but to do no real harm (other than to your wallet).

An excellent article entitled, "A Homeowner's Guide to Septic Systems" (Sponenberg et al., 1985, p. 8), identifies a series of dangerous chemicals, stating, Do not use products containing more than one percent by weight of the following chemicals:

Halogenated Hydrocarbons

Trichloroethane
Trichloroethylene
Methylene chloride
Halogenated benzenes
Carbon tetrachloride

Aromatic Hydrocarbons

Benzene
Toluene
Napthalene

Phenol Derivatives

Trichlorophenol
Pentachlorophenol
Acrolein
Acrylonitrite
Benzidine

I have been bothered for some time by the number of experts who flatly rule out the use of any additives, for two reasons. First, while I am not a specialist in the field, I do remember from my undergraduate courses in sewage treatment and water supply that chemicals are commonly used: chlorine, florides, activated charcoal, to recall a few. So why, I mused, should *all* additives be summarily ruled out in the companion field of septic system technology? Second, except for certain moral issues and missing putts less than two feet, I have always subscribed philosophically to the adage, "Never say never." Thus, I am in accord with the position of the EPA, as stated by Jim Kreissl, one of their environmental engineers; "The EPA's position on additives is that few have shown any benefits and some chemical additives have been proven to have negative effects, such as contamination of the groundwater. [However], we like to leave the door open to any manufacturer who wants to prove their product is effective."

I have a suggestion. A few years ago, I read an article in the newspaper about a bacterial agent that was going to be used to attempt to clean up an oil spill off the coast of Galveston, Texas. The bacteria, it was said, "eats oil." By then well along in my private study of septic system technology, and being aware of how detrimental oil in the septic tank could be, I thought, "Why not use this stuff in septic systems?" Subsequently, I was invited to do a paper at a water resources conference (Monahan, 1990), at which I broached the idea to the audience. Afterward, I was approached by an engineer from the Department of Environmental Protection of New York City, Joseph P. Conway, P. E., who liked my idea, and said that New York City had been using just such a bacteria with great success in their problem sewers. He promised to send me some information from their files. He did so, and the evidence and documentation was extensive and impressive. I thank Mr. Conroy for sharing the information.

The material is called DBC-Plus. The DBC signifies dried bacteria cultures. Initial use appears to have been in New York City to clear sewer lines clogged with grease, notably in the areas of Chinatown and Restaurant Row where oils and greases from restaurants created a special problem of long duration. As reported in the *Daily News* (5 September 1982) and *Engineering News Record* (9 September 1982), the dry powder is mixed with warm water and "poured into a blocked sewer and in about two days the bacteria consume enough grease to separate the gunk from the sewer walls. A log-shaped mass of the gunk is then flushed down to the next manhole and removed" (*Daily News*).

Mr. Conway also included copies of letters from four New Jersey towns, all of which had similar success using the material. Fort Dix, NJ, also had success, and also reported favorable evaluations from the army's Construction Engineering Research Laboratory in Champaign, IL. A paper was included that briefly described a case history of interest:

> The commercial use of bacterial formulations in dealing with petroleum wastes has been done for many years. Back in 1968 the *Queen Mary* was dry docked to be converted to a museum. Only waters within the bilges posed a serious fire hazard for using the acetylene torches necessary to cut out the some 8,000 tons of machinery that needed to be removed.
>
> A commercial bacteria formulation, Type L, DBC Plus, was used. Within 48 hours the Long Beach Fire Department approved the use of torches so that the engine room project could begin (Grubbs, 1983).

One additional enclosure from Conroy, which appears to be a commercial flyer from the manufacturer, lists some intriguing additional benefits of DBC Plus, namely that it is "made up of harmless saprophytes" and thus is "safe to use," and is "not detrimental to plants or wildlife, nor will it damage pipes or equipment." It also "improves BOD_5 and suspended solids removal, and lowers sludge volumes."

The flyer ends with the statement that DBC Plus "restored percolation in leach fields." No further amplification of this statement was included, but what is clearly implied is that the material has already been used successfully as a septic system additive.

What is surprising then, is the fact that nowhere else have I seen any reference to this material, or even to the general idea of using "oil-eating" bacteria as an additive. Hopefully, someone will check this out, notably the EPA people who assert that they are open to new possibilities.

The material is made by Flow Labs Inc. of Inglewood, CA, and is distributed by M.L. Gold and Associates, Inc., P. O. Box 2134, Cherry Hill, NJ 08034. Phone (800) 422-4468.

One final caveat to owners/consumers who are using, or are considering using, some advertised additive: Insist on some independent verification of claims made by the manufacturer, and be certain that the verification comes from a reputable source that has no axe to grind. A number of organizations are springing up that test purported "green" products for environmental friendliness, some academic and some professional. *Consumer's Digest*, for example, accepts no advertising, and is highly regarded for their reliability and independence.

With all of the extraordinary additional complications put before the con sumer by the dizzying array of additives available, with some experts saying the additives are useless, and manufacturers claiming to have the "magic bullet," it becomes something of a riddle wrapped within an enigma to figure out what to do. In some jurisdictions, a process of septic system management has emerged. To conclude this chapter, following is a brief description of how this has worked with a considerable degree of success.

9.6 SEPTIC SYSTEM MANAGEMENT

About 20 years ago, some small, rural communities where conventional sewered systems were not economically feasible started to realize that the common practice of simply turning over the septic system to the homeowner for management and maintenance was ill-advised, and that professional management of groups of septic systems made more sense. The Georgetown Divide Public Utility District in California is generally recognized as the pioneer in onsite management, starting in the mid 1970s. Since then, a variety of new management agencies have evolved, with water utilities leading the way. The *Small Flows* issue of July 1992 describes a number of such efforts in California, Missouri, Maryland, and Michigan. After 21 years of operation and maintenance of 740 systems, the California utility has experienced few system malfunctions. The costs of the services seem to be quite nominal and remarkably similar in the four widely separated jurisdictions, generally ranging from about $12 to about $20 per month.

The process involves all of the steps discussed in this chapter, from site evaluation to *communication with the homeowners after the systems are built.* The intermediate steps of design and construction are carefully planned and monitored.

As I read the descriptions in the series of articles, I was struck with the common theme of constructive interaction between the managing agencies and the owners. Indeed, in one case, it was the developer who invited the agency to assume the role of management. Such phrases as "community outreach" and "public cooperation" were common in the descriptions.

Contrast this to the tone of the newspaper article quoted earlier in this chapter, illustrating the confrontational, punitive, and almost vengeful attitude of Pennsylvania's DER, who continually provide evidence of this negative and self-defeating attitude toward the public. Thus, while I endorse the concept of professional septic system management, I believe that it is vital that the agency that does the managing exhibit the kinds of positive and helpful attitudes so evident in the descriptions of the systems in the *Small Flows* articles.

The idea that I have stressed repeatedly about disseminating useful information to the public (especially concerning maintenance) is in no way diminished by the concept of professional management. Indeed, the two approaches are complementary, since it is the homeowner who manages his or her system on a daily basis. If done well, the *physical* inspections and maintenance by the professionals will be simple and straightforward, and will, I think, result in relatively trouble-free decentralized sewerage systems wherever they are installed, with a consequent benefit to the overall environment at the least cost and aggravation for all concerned.

9.7 GLOSSARY

Evapotranspiration The natural process whereby plants remove water from soil through their root systems, branches, and leaves, with the water entering the atmosphere through evaporation of the vapor from exposed surfaces.

CHAPTER 10

Techonomics

Having discovered the fun of inventing new words early in this work (gemicoss), I thought I'd try again with the title of this short chapter. It is short because it is a subject about which I know very little, largely because I have spent most of my professional career in the academic world where one does not deal very much with the business and economic aspects of consulting, construction, and interaction with clients regarding business economics. I will also say that I have always had a sensitivity toward, and appreciation for, the important link between technology and economics.

What I have to say comes from this interest, but is largely supported and augmented by conversations that I have had on these subjects with professional colleagues.

10.1 ENGINEERING DESIGN

The conversation I had that prompted me to think about adding this chapter was with an executive of Woodward–Clyde Consultants (geotechnical specialists). By interesting coincidence, this firm is the same one—now grown considerably larger—where I had my start in professional geotechnical engineering, and where almost all of my early experience and interest in compacted fills was obtained. So it is, I think, quite fitting that this chapter came about in that way, sort of a closing of the cycle. In this conversation a concern was expressed that too many engineers were not sufficiently sensitive to how economics *should* influence technical decisions. A specific case history was cited as an illustration.

A foundation investigation was made for a warehouse for a large greeting card company. The warehouse site was in a location where the subsurface conditions were very poor, with soft compressible soils to substantial depth. After appropriate sampling, testing, and analyses, the engineers recommended the site stabilization technique known as "surcharging with sand drains," a now-common procedure where vertical sand columns (or, more recently, artificial "wicks") are installed in a designed grid pattern to appropriate depths, and a surcharge load is placed on the site to induce drainage (preconsolidation). The load is created by a fill of specific design thickness. The process results in settlement and increased strength of the subsurface soil. The typical design time for such stabilization to occur is about one year. The surcharge is then removed (usually) and the structure is built on the preconsolidated, stabilized site.

As was pointed out, however, while the design was technologically correct, it was done without any economic considerations. The "cost of money," as it was phrased, was about 15%, and the one-year period would also deprive the client of the use of the warehouse for a full December holiday season, a period when this particular client's profits are obviously concentrated.

Until and if interest rates come down significantly, it would appear that the surcharging procedure might not be viable for any projects, certainly for a lot fewer than in the old days of 6% interest rates. The second consideration is, of course, unique to this client. Such considerations dictated a pile foundation, in that the interest cost would be largely eliminated and the warehouse would be completed for use by the holiday period. Since I had this conversation, it has occurred to me that the use of lightweight fill in a weight credit application might have proved to be a viable alternative consideration, as was described in Section 5.3.1.

10.2 LEGAL COSTS

Another aspect of techonomics that deserves comment is legal costs. I have discussed some of the issues with colleagues and done additional research to further educate myself for the preparation of what follows. Actually, long before I contracted to write this book, I was interested in how legal matters affected my personal economics in terms of part-time consulting activity. So my studies precede research for this book.

The nation's media deal increasingly with the burgeoning problem of the growing number of lawsuits and their cost. Most recently, Herb Jaffe (*The Star Ledger*, Newark, NJ, 22 January 1985) referred to "a steadily rising lawyer population . . . new laws [generating] lawsuits" contributing to "the litigation nightmare of the 1980's." One statistic that suggests the proportions of the problem is that more than 40,000 new lawyers a year graduate from 175 accredited law schools in the United States. Increasingly larger judgments, lawyer's fees, expert witness fees, and court costs, combined with the much larger number of lawsuits, drives the cost of professional liability insurance to prohibitive levels.

Attorney's fees and expert witness fees are often very high. In a recent newspaper description of the settlement of the Agent Orange case, one lawyer was reported to have requested the eye-opening rate of $450 *per hour*. Expert witness fees, while usually not in that range, can also be impressive. Such fees are typically substantially higher than hourly rates charged for standard engineering design services. I have often thought it anomalous that this is so. It has always seemed to me that it is easier to figure out what went wrong afterward than to avoid error in design or construction. On top of that, there's less risk! While I suppose it happens, how often do you hear about an expert witness getting sued? To make it more anomalous, the witness doesn't even have to be right, he just has to make the most convincing case.

I recall an interesting and fascinating case in point. Some years ago, G. A. Leonards, of Purdue University, delivered the annual Terzaghi Lecture, which dealt with a major pile foundation failure (in Africa, as I recall), for which he was the eighth of a succession of experts. I vividly remember Professor Leonards' last slide, a summary of the eight explanations of the cause of the failure, all of which were (I recall) mutually exclusive.

Thus, a case can be made that the costliest engineering work is easier, less risky, and you don't even have to be right. (In fairness, perhaps it should be added that the *pressures* of such work can be substantial—such as eight hours of hostile questioning.)

As other observers have noted, legal costs are passed on to clients. It is also likely that direct engineering costs are also higher because of an understandable tendency to do more rather than less sampling and testing because of the fear of lawsuits, rather than for strictly technological reasons.

10.2.1 A Proposal

I have a specific suggestion for at least a partial solution: a concerted effort directed toward the elimination or substantial reduction of the attorney's contingency fee system in the United States. What prompted this idea was an interesting and intriguing discovery that this system is not used in Canada. I learned this from a document that was loaned to me by a colleague with concerns on the subject similar to mine, Ed Dauenheimer of the New Jersey Institute of Technology. Actually, his concerns are greater than mine, since he is a specialist in construction planning, having held professional engineer's licenses in 18 states, with extensive experience in engineering construction practice, and presently engaged in part-time consulting. The document is a pamphlet entitled, "Professional Liability Loss Control: Architects and Engineers." It is printed and distributed by INAX Underwriters Agency, Inc.,

Chicago, IL, a company that advises clients on how to minimize their exposure and losses from lawsuits.

In discussions with Professor Dauenheimer I learned that concerns about the costs of potential lawsuits have become so pervasive that (1) liability insurors actually dictate to potential clients what language, and even words, may and may not be used in contract documents, or else they will not provide coverage (maybe that's how "observers" originated), and (2) INAX provides clients with a thick looseleaf binder, mostly composed of *do's* and *don'ts*; according to Professor Dauenheimer, "mostly Do Not's."

In my experience as an expert, the opposing lawyer always asks, "How are you compensated for your services?" and I reply, "On an hourly fee basis." The implication is clear: If you're paid on a contingency basis, your testimony will be portrayed as biased because of your financial stake in the outcome. I have never understood why the same logic does not apply to attorneys.

The contingency fee system, as stated in the INAX pamphlet, allows a plaintiff to sue with little or no expense. Thus, the plaintiff has everything to gain and nothing to lose by suing, and it is becoming more and more common for lawyers to advertise their services using the contingency feature as the principal focus of the advertisement. The INAX pamphlet further states that "conversely, in Canada, a plaintiff must assume a substantial expense in filing a suit, since the attorney is paid a flat fee or retainer. . . ."

What specific steps can be taken to eliminate or modify the system? First, I'd suggest that a coalition of some sort is needed, one with much clout, since lawyer's groups would doubtlessly provide formidable resistance. I hope, however, that many lawyers and judges would agree with at least some of the arguments presented here. The coalition could include professionals from The American Society of Civil Engineers, The National Society of Professional Engineers, The American Institute of Architects, and contractors' associations. Because of the broad national scope of the problem it might be feasible, and it certainly would introduce more clout, to attempt to include the Ralph Nader organization.

Second, find out how Canada does it. Since Canada is a free society like ours, I presume it's not through government control. So it is likely that a self-regulatory body is responsible. If that is the case, that information could be used as pressure to reform the American system.

Perhaps some intermediate solution is possible and desirable, wherein a system could be developed by the American Bar Association to impose limits of some sensible dimension on its members, much like the ban on advertising (recently relaxed). This might be desirable because of the valid argument that people of limited means would have no legal recourse without a contingency system. Perhaps state boards could be established to consider applications from aggrieved parties who could not otherwise afford to sue. Again, what does Canada do in this regard?

It would be interesting to compare the claims statistics of the United States and Canada for the same time period, assuming the data for the latter could be obtained. (The U.S. statistics for most of the decade of the 1970s are presented in chart form in the INAX pamphlet.) Perhaps that would provide some indication of the effects of the contingency system. A valid and meaningful study of such a complex problem would need to be much more comprehensive, since many variables other than contingency fees are involved, but this would be a start. This would be a great topic for a social science major.

CHAPTER 11

Appendices

This final chapter includes sections on very large jobs, and one entitled Annotated References for Further Study. There appears a list of references that were used (and cited within the text) for the preparation of the book. It has occurred to me, however, that some additional references, not cited in the book, would be helpful to readers wishing further information on specific subject matter not fully covered here. Field load testing, for example, is mentioned only peripherally in the text, since a full treatment would not be consistent with the book's focus. Nonetheless, since one corollary purpose of the book is to provide what amounts to a field manual (notably Chapter 8), the same people interested in the field work described would also be interested in all other aspects of field work. The self-study references (including some excellent sound film strips) are listed here to accommodate these people and their busy supervisors.

Finally, a section describing available seminars and lectures is included.

11.1 VERY LARGE JOBS

With the exception of very brief descriptions (subtitled "Specification Components") in Section 7.1.1., the problems that are characteristic of very large compacted fill jobs have not been included in this book, for three reasons. First, I have endeavored to focus on describing first-hand experiences, and mine have been limited to small jobs. (As will be explained, "small," and "large," and "very large" are very subjective descriptive terms.) Second, the book is directed mostly to nonspecialists and aspiring young and inexperienced geotechnicians. Finally, it is inconceivable that jobs of the magnitude I am about to describe would (or could) be considered without the careful and expert input of geotechnical specialists.

Because of these reasons, I will provide only a summary description of the project highlights. Fortunately, an excellent paper has been published, and was recently brought to my attention by its senior author, Dennis J. Leary, P.E., who served as the project engineer. The paper is entitled "Earthwork Quality Control for Ludington Pumped Storage Reservoir," published by the University of Wisconsin—Milwaukee and the American Water Resources Association. The paper was presented at the International Conference on Pumped Storage Development and Its Environmental Effects, Milwaukee, Wisconsin, 1971.

The project involved the construction of a six-mile-long earth embankment (the upper reservoir) on the east shore of Lake Michigan, which included five major embankment soils (textures) from a common borrow. The total volume of the embankment was 39 million cubic yards, placed at rates as high as 180,000 yards per day. (Contrast this to Case Study 1, Section 4.1.1, where the *total* fill volume was 150,000 yards.) In addition, a compacted clay impervious lining, 5–8 ft thick, was constructed. Filter sands were hauled approximately 40 miles by truck and base drain gravel was imported from the Upper Peninsula at the rate of three ships per week.

Two teams were involved with the work: a quality control inspection staff composed of a resident engineer, two field engineers, and 26 inspectors; and a quality control field and laboratory testing staff composed of the soil engineer, 4 engineers, and 14 field and laboratory technicians. The latter team was divided, interchangeably and flexibly, into three groups: field testing, laboratory testing, and data processing.

Methods specifications were used, including the use of more than 50 vibratory sheepsfoot compactors, rubber-tired tractors, and static sheepsfoot rollers (for the clay lining). Loose lift thicknesses, coverages (passes), maximum roller speeds, and water contents were stipulated for each soil type.

Field tests included in-place density tests (Washington Dens-O-Meter), loose lift thickness measurements, and static cone penetrometer soundings. The latter tests were done to evaluate and thus ensure uniformity of compaction, augmenting the field-density test results.

Visual and manual soil descriptions and classifications were made by field technicians and verified by laboratory tests. Index property tests included liquid and plastic limits and grain size distribution determinations. Engineering tests included the standard Proctor compaction test and the relative density test (depending on fines content), and permeability tests of the various embankment zone materials.

As may be inferred by the magnitude and scope of this work, the principal distinctive feature of such jobs may be summed up in one word: management. Management of the personnel and logistics that are involved, and management of the enormous amounts of data that are generated. Implicit in management of personnel and logistics is the setting up of organizational charts, modes of communication (especially systematic reporting), the training of individuals, and the organization of individuals into working teams. The analysis of test results, especially where voluminous data sets are involved, is increasingly being done by statistical methods, and this was done for the Ludington project.

Methods of determining and ensuring representivity of test results were done in an interesting and perhaps unique way: "Density tests were required to meet criteria concerning depth, distance from unconfined slopes, and soil conditions."

As stated by the authors, "tests made too near the surface give low and unrepresentative results." This observation corresponds with my commentary about density gradients and their investigation (Section 5.5), and the need to "prepare a fresh surface of the compacted fill . . . by scraping off the top few inches of the rolled fill to eliminate effects of shrinkage by desiccation [and other surface disturbance]" (Section 8.1.2). I was interested to learn that these effects were taken into consideration in the Ludington project. More interesting is the fact that a method was developed (but not explained) for choosing a specific test depth for different soil types (varying from 0.5 to 2.0 ft).

The requirement for "minimum distance from unconfined slopes" (for the location of field-density test holes) is, to my knowledge, unique. It is a very logical requirement. (In fact, when I read the paper, I thought, "Why didn't I think of that?") For the Ludington project, the specified minimum was 5 ft from unconfined slopes, and 10 ft from "unconfined corners" (e.g., in sand blanket near the intersection of longitudinal and lateral [orthogonal] drains). The authors state that "these distances were established on the basis of test results obtained at the start of embankment construction." Should readers wish to obtain further information regarding these criteria, or other aspects of the Ludington project not covered in the published paper, contact Dennis J. Leary, P.E., Langan Engineering Associates, Inc., River Drive Center 2, Elmwood Park, NJ 07407.

The third criterion, soil conditions, is directed to the not uncommon occurrence where the texture of the soil removed from the density test hole differs significantly from that of the laboratory sample that was used for the compaction test. The authors state that (in such a case) "the hole is abandoned." This is in accordance with commentary I have made with respect to simulating field conditions (Section 8.1.2). Unfortunately, neither Leary nor Monahan have established any quantitative criteria for determining, with any meaningful exactness, *when* to abandon the hole.

11.2 ANNOTATED REFERENCES FOR FURTHER STUDY

To provide additional help to those starting in any of the various aspects of engineered construction, I recommend the following for self-study. The acquisition of these books and audiovisual materials by employers will also serve to minimize the time and expense for supervisory and training activity.

One caution, however: Review the materials. While everything I list here is generally recommended, I do have some reservations about some of the advice given. There are some things in some Soiltest filmstrips with which I disagree, for example. A suggested approach is to view the filmstrips initially with the staff people, review the work afterward, and perhaps prepare a written critique detailing differences, additions, or corrections. Such a document could then be distributed to future viewers (new staff) as an addendum to the filmstrip.

11.2.1 Soiltest Filmstrips

As a part of the laboratory component of the undergraduate course in soil mechanics that I developed and taught for over 26 years at Newark College of Engineering (NCE), these Soiltest sound filmstrips were required viewing during periodic scheduled laboratory write-up sessions:

1. Auger borings (including hollow stem)
2. Split-barrel sampling/standard penetration test
3. Thin-walled tube sampling (shelby tube)
4. Field-density testing—sand cone method

Each filmstrip runs about 20 to 25 minutes. The filmstrips were chosen to allow some coverage of important field operations necessary to geotechnical work. Prior to this solution, I wrestled with the problem of how to get this type of information across to the students, since I judged that it would be inappropriate to spend significant amounts of time on such subject matter in classroom lectures, where soil mechanics theory was, appropriately, of highest priority. Geotechnical specialists will recognize the importance of students gaining at least some knowledge about the operations of the filmstrips listed, but it would be beyond the scope of this book for me to include anything but the most general observations about the content, since the focus of this chapter is self-study.

Two additional filmstrips that I showed in the laboratory course were "Atterberg Limits Testing" and "Unconfined Compression Testing." These were shown in the early part of the course, after orientation sessions and some hands-on experience in textural classification (grain size distribution by visual and tactile inspection and by sieving, and subjective evaluation of plasticity of cohesive soils, all by Burmister's field inspection techniques). For efficiency and convenience, these two filmstrips were scheduled for viewing by the entire class on the day scheduled for the respective tests. As I recommend to supervisors, I would follow up the filmstrip viewing by about a 15- to 20-minute commentary on what the students had just seen, thus modifying the content of the filmstrips to my own purposes. The students scheduled for the test that day would then move to the adjacent laboratory to perform the test, with the filmstrip and my related commentary fresh in mind. (Preparatory to the laboratory, they would—or should—have read the appropriate sections of the laboratory manual.) This time-tested procedure worked quite well in an academic setting, and I would judge it would also work in training people in an in-house, professional setting.

Now, a few editorial observations about the Soiltest filmstrips. I believe, in general, that the producers have let their zeal to sell equipment affect their directorial judgment. They cater to the very lowest of intelligence, I guess figuring that they'll have a broader market. In one frame, for example, they say, "This is a knife."

In some filmstrips, there are out-and-out mistakes, or at the least, poor advice. In the "Unconfined Compression Test" filmstrip, the narrator stumbles over a dial reading, and there is a strange and repetitious emphasis: "Be sure the strain dial is set to zero," an arbitrary and unnecessary requirement. Conversely, they do not mention the real need to be certain of *positive contact* of the strain apparatus at the initiation of load application.

Soiltest filmstrips can be a worthwhile investment, as useful teaching aids, if supplemented by appropriate review and commentary from experienced supervisors. A little humor can be helpful in softening the excesses of the commercialism. Soiltest has more than 20 filmstrips available, ranging from simple laboratory index testing techniques to the use of sophisticated instrumentation for seismic and resistivity field testing. For the more complex subject matter, such as the latter, they also offer geophysical training programs. Their corporate headquarters is Soiltest, Inc., 2205 Lee Street, Evanston, IL 60202.

11.2.2 Encyclopaedia Britannica Filmstrips

Two of the finest sound filmstrips I have seen are from the Basic Earth Science Program, coproduced by Encyclopaedia Britannica and the American Geologic Institute, copyright 1969:

1. "Geologic Measurements and Maps," Series No. 6414
2. "Investigating Rocks," Series No. 6415

They are recommended without reservation as effective self-study materials. These filmstrips (and probably many others) are available from Encyclopaedia Britannica, 425 North Michigan Avenue, Chicago, IL 60604, or 175 Holiday Inn Drive, Cambridge, Ontario, Canada N3C3N4.

11.2.3 Miscellaneous References

1. *Soil Mechanics Laboratory: Procedures and Write-up*, E. J. Monahan, Ph.D., P.E. I have available an unpublished 90-page manual that I developed for the

course I taught at NCE. It is elementary and presumes no knowledge or experience in soils. People developing such courses at the growing number of community and county colleges, particularly BSET programs (Bachelor of Science in Engineering Technology), may find it helpful, since scheduling and a "To the Instructor" section are included, detailing (for example) an hour-by-hour description for the early (orientation) weeks of the course.

A variety of readers may be interested in the detailed treatment of procedures for soil identification and classification. Suggestions regarding report writing are also included. The manual is in printed form, but not of polished, published format.

2. *Basic Procedures for Soil Sampling and Core Drilling*, W. L. Acker III. This highly recommended document is distributed by the Acker Drill Co., P.O. Box 830, Scranton, PA 18501.

3. *The Design of Foundations for Buildings*, Johnson and Kavanaugh, McGraw-Hill, 1968. This reference is cited in my reference list, but for another reason. I mentioned in the text that detailed descriptions of load testing were beyond the scope of this book. Johnson and Kavanaugh have an excellent treatment of this subject in Chapter 9.

4. *The Effects of Angularity on the Compaction and Shear Strength of a Cohesionless Material*, Richard Swiderski, master's thesis, Newark College of Engineering, 1976. This master of science thesis on angularity and its effects on soil behavior is listed because I think the topic is important and has not been given the attention it deserves. The investigation was the combined effort of two excellent students, one undergraduate senior civil engineering student (Jeff Tubelo), and the author, a graduate student in civil engineering with an undergraduate degree in geology. The different but related undergraduate backgrounds contributed to an interesting and rewarding study.

11.3 SEMINARS AND LECTURES

As indicated in the preface, F. C. Budinger has available an excellent and highly recommended two-day seminar, the parts of which are "Understanding Soils," "Measuring Compaction," "Constructing with Earth," and "Testing Fills." He has regularly scheduled individual registration seminars (mostly in the West), and is also available for in-house seminars, by arrangement. To get on his mailing list, or to contact him about an in-house seminar, write to Bender & Associates, E. 3820 Broadway, Spokane, WA 99202. Phone (509) 534-1426; Fax (509) 535-9589.

And if you're interested in a lecture *and* a story, contact me at 85 Newark Avenue, Bloomfield, NJ 07003. Phone (201) 743-6210/6043.

REFERENCES

Ahmed, I. and C. W. Lovell, "Rubber-Soils as Lightweight Geomaterials," Transportation Research Board, 72nd Annual Meeting, Washington, DC, 10–14 January 1993.

Allen, T. M. and A. P. Killian, "Use of Wood Fibre and Geotextile Reinforcement to Build an Embankment Across Soft Ground," Transportation Research Board, 72nd Annual Meeting, Washington, DC, 10–14 January 1993.

Baker, L. K., et al., "Household Water Conservation Effects on Water, Energy and Wastewater Management," The Second Annual Conference on Complete Water Use, Chicago, IL May 1975.

Breed, C. B., and G. L. Hosmer, *Elementary Surveying*, Vol. 1, 8th ed., Wiley, New York, 1945.

Brown, V. J., "Soil Compaction in Narrow Places," Arrow Manufacturing Co., Denver, CO, 1967.

Burmister, D. M., "Identification and Classification of Soils," ASTM STP No. 113, Am. Soc. Testing Mats., pp. 3–18, 1953.

Burmister, D. M., *Soil Mechanics*, Vol. I, Columbia University Press, New York, 1955.

Burmister, D. M., "Suggested Methods of Test for Identification of Soils," *ASTM: Procedures for Testing Soils*, Philadelphia, 1958.

Carson, A. Brinton, *General Excavation Methods*, Krieger, Huntington, NY, 1980.

Casagrande, A., "Seepage Through Dams," *Contributions to Soil Mechanics: 1925–1940*, Boston Society of Civil Engineers, 1940, p. 295.

Cedergren, H. R., *Seepage, Drainage, and Flow Nets*, 3d ed. Wiley, New York, 1990.

Chae, Y. S., and T. J. Gurdziel, "New Jersey Fly Ash as Structural Fill," *New Horizons in Construction Materials*, Vol. I, Envo Publ., Lehigh Valley, PA, 1976.

Coleman, T. A., "Polystyrene Foam is Competitive, Lightweight Fill," *Civil Engineering*, February 1974, pp. 68–69.

Collier, C., "Wetlands: Untangling the Issues and Regulations," WRA/DRB 30th Annual Spring Conference, 13 April, 1989, Cherry Hill, NJ, WRA/DRB, Valley Forge, PA, p. 30.

Converse, J. C., et al., "Design and Construction Procedures for Mounds in Slowly Permeable Soils with Seasonally High Water Tables," Small Scale Waste Management Project, University of Wisconsin, Madison, WI, 1975.

Converse, J. C., "Publication List of the Small Scale Waste Management Project," Department of Agricultural Engineering, University of Wisconsin, Madison, WI.

Conway, J. P., P.E., Director, Department of Environmental Protection, City of New York. Personal correspondence, November 1990.

deBoer, L., "Expanded Polystyrene in Highway Embankments," *Geotechnical News*, March 1988, p. 25.

"Domestic Policy Council's Task Force on Wetlands: Summary of Public Meetings and Written Comments," February 1991, U.S. Department of the Interior, Washington, DC, NTIS PB81-161182.

Dragan, J. (ed.), *Journal of Soil Contamination*, Lewis Publishers, Boca Raton, FL.

Edil, T. B., and P. J. Bosscher, "Development of Engineering Criteria for Shredded Waste Tires in Highway Applications," Final Report, Research Report Number WI 14-92, Department of Civil and Environmental Engineering, University of Wisconsin, September 1992.

Eggestad, A., "Experience of Compaction Control in Sand and Gravel," *International Conference on Compaction*, Vol. II, Paris, 1980.

Elastizell Corporation technical literature, Ann Arbor, MI, 1993.

ENR, "Plastics Replace Subsoil," *Engineering News Record*, 27 April 1989, p. 17.

Fang, H. (ed.), *New Horizons in Construction Materials*, Vol. I, Envo Publ., Lehigh Valley, PA, 1976.

Flaate, K., "Superlight Material in Heavy Construction," *Geotechnical News*, Vol. 5, No. 3, September 1987, pp. 22–23.

Flaate, K., "The (Geo)Technique of Superlight Materials," *The Art and Science of Geotechnical Engineering: At the Dawn of the Twenty-First Century, A Volume Honoring Ralph B. Peck*, Prentice-Hall, Englewood Cliffs, NJ, 1989, pp. 193–205.

Fletcher, G. A. and V. A. Smoots, *Construction Guide for Soils and Foundations*, Wiley, New York, 1974.

Frydenlund, T. E., "Expanded Polystyrene: A Lighter Way Across Soft Ground," Intern rapport nr. 1502, Veglaboratoriet, Oslo, Norway, May 1991.

Giddings, T. R., "A Rapid Method of Controlling Compaction by Plate Loading Tests," *International Conference on Compaction*, Vol. II, Paris, 1980.

Greer, D. M. and D. C. Moorhouse, "Engineering—Geologic Studies for Sewer Projects," *Journal of the Sanitary Engineering Division, Proceedings, ASCE*, February 1968.

Grubbs, R. B. "Jones," "Environmental Applications of Biotechnology: The Current State of the Art," Genetic Control of Environmental Pollutants Conference, University of Washington, Seattle, August 1983.

Healy, M., "Utilization of Marginal Lands," Master's project (unpublished), New Jersey Institute of Technology, Newark, NJ, 1975.

Hendron, D. M. and L. Holish, "Quality Control of Earthwork Construction Using Cohesive Soils with Highly Variable Properties," *International Conference on Compaction*, Vol. II, Paris, 1980.

Hirschfeld, R. C., "Shear Strengths for Analysis of the Stability of Earth-dam Slopes," American Society of Civil Engineers Seminar, New York, 2 June, 1965.

Horvath, J. S. (1992b), "Dark, No Sugar: A Well-known Material Enters the Geosynthetic Mainstream," *Geotechnical Fabrics Report*, October 1992.

Horvath, J. S., " 'Lite' Products Come of Age: New Deelopments in Geosynthetics," *ASTM Standardization News*, September 1992, pp. 50–53.

Hough, B. K., *Basic Soils Engineering*, 2nd ed., Ronald, New York, 1969.

Humphrey, D. N., T. C. Sandford, M. M. Cribbs, and W. P. Manion, "Shear Strength and Compressibility of Tire Chips for Use as Retaining Wall Backfill," Transportation Research Board, 72nd Annual Meeting, Washington, DC, 10–14 January 1993.

Johnson, S. M. and T. Kavanaugh, *The Design of Foundations for Buildings*, McGraw-Hill, New York, 1968.

Kaplan, O. B., *Septic Systems Handbook*, 2d ed., Lewis, Chelsea, MI, 1991.

Kleckner, D., "Bush Announces Wetlands Policy," *Pocono Record*, 10 August, 1991.

Koerner, R. M. and J. P. Welsh, *Construction and Geotechnical Engineering Using Synthetic Fabrics*, Wiley, New York, 1980.

Laguros, J. and J. Robertson, "Compaction of Cover Material in Refuse Landfills," *International Conference on Compaction*, Vol. I, Paris, 1980.

Lambe, T. W., *Soil Testing for Engineers*, Wiley, New York, 1951.

Lambe, T. W., "The Engineering Behavior of Compacted Clay," *Journal of the Soil Mechanics and Foundations Division, Proceedings, ASCE*, May 1958.

Leonards, G. A., "Pavement Construction," *U.S. Patent 3,250,188*, 10 May, 1966.

Lowrance, W. W., *Of Acceptable Risk*, Kaufmann, Los Altos, CA, 1976.

Maine State Highway Commission, "Insulation of Subgrade—Evaluation of First Year Data," Dow Chenmical Co., Midland, MI, 1966.

Mardekian, A., W. Rowbotham, and B. Facente, "Soil Compaction Comparative Studies," Senior project (unpublished), New Jersey Institute of Technology, Newark, NJ, 1973.

May, R., "The Infiltrator Leaching System," Technical Support Paper, Infiltrator Systems, Inc., Old Saybrook, CT, 1989.

McBee, W. C., D. Saylak, T. A. Sullivan, and R. W. Barnett, "Sulfur as a Partial Replacement for Asphalt in Bituminous Pavements," *New Horizons in Construction*, Vol. I, Envo Publ., Lehigh Valley, PA, 1976.

Means, R. E. and J. V. Parcher, *Physical Properties of Soils*, Merrill, Columbus, OH, 1963.

Meehan, R. L., "The Uselessness of Elephants in Compacting Fill," *Canadian Geotechnical Journal*, September 1967, pp. 358–360.

Modern Plastics Encyclopedia, McGraw-Hill, New York, 1968.

Monahan, E. J. Floating Foundation and Process Therefor, *U.S. Patent 3,626,702*, 14 December 1971.

Monahan, E. J., Novel Low Pressure Back-Fill and Process Therefor, *U.S. Patent 3,747,353*, 24 July, 1973.

Monahan, E. J., "A Method for Specifying Percentage Soil Compaction," *Civil Engineering*, May 1974, pp. 68–69.

Monahan, E. J., "Septic Systems Technology," Conference of The Water Resources Association of The Delaware River Basin, White Haven, PA, October 1990.

Monahan, E. J., "Weight-Credit Foundation Construction Using Artificial Fills," Transportation Research Board, 72nd Annual Meeting, Washington, DC, 10–14 January 1993.

Morris, M. D., "Earth Compaction," *Cosntruction Methods and Equipment*, McGraw-Hill Reprint, New York, 1959.

Negussey D. and M. Jahanandish, "A Comparison of Some Engineering Properties of EPS to Soils," Transportation Research Board, 72nd Annual Meeting, Washington, DC, 10–14 January 1993.

Peck, R., "Art and Science in Subsurface Engineering," *Geotechnique, International Journal of Soil Mechanics and Foundations*, March 1962.

Peck, R. and H. Ireland, "Backfill Guide," American Society of Civil Engineers meeting, Jackson, MS, 1957.

Peck, R. B., W. Hanson, and T. Thornburn, *Foundation Engineering*, Wiley, New York, 1953.

Perkins, R. J., *Onsite Wastewater Disposal*, Lewis, Chelsea, MI, 1989.

Pocono Record, "Bush Announces Wetlands Policy," 10 August 1991.

Pocono Record, "Sewage Inspection Proposal Rapped as Too Far-Reaching," 26 March 1993.

PP&L, "Storage Heating: A Good Idea for Old and New Homes," Pennsylvania Power and Light brochure, Allentown, PA, January 1988.

Proctor, R. R., "Fundamental Principles of Soil Compaction," *Engineering News-Record*, 31 August, 9 September, 21 September, 28 September 1933.

"Professional Liability Loss Control: Architects and Engineers," INAX Underwriters Agency Inc., Chicago, IL, 1980.

Public Utility Commission, Pennsylvania, "Saving Water Around the House," April 1991.

Public Works, "Plastics Recycling at Record Rates, But Markets Still Lag," July 1992, p. 7.

Reichert, J., "Various National Specifications on Control of Compaction," *International Conference on Compaction*, Vol. III, Paris, 1980.

Richards, F., *Sue the Bastards: Handbook for the Field Engineer*, Richards, Pittsford, NY, 1976.

Ritter L. J. and R. Paquette, *Highway Engineering*, 2d ed., Ronald, New York, 1960.

Rotordisk, Ontario, Canada, technical literature, 1993.

Ryan, J. D., Consulting Geologist, "Examination of Soil Profiles on Wagner Forest Park, Tobyhanna Township, PA," January 1976.

Schmertmann, J. H. and D. K. Crapps, Various reports, Gainesville, FL, 1986.

Schroeder, W. L., *Soils in Construction*, 2d ed., Wiley, New York, 1980.

"Sewage Facilities Act," 24 January 1966, and related Rules and Regulations: Title 25, Pennsylvania Code, Chapters 71 and 73.

Small Flows, July 1992, West Virginia University, Morgantown, WV.

"Small Wastewater Systems: Alternative Systems for Small Communities and Rural Areas," U.S. EPA, January 1980, FRD-10.

Smith, T., CMS Rotordisk, Ontario, Canada, personal correspondence, 1993.

Solite Corporation technical literature, Mt. Marion, NY, 1993.

Sowers, G. F., *Introduction to Soil Mechanics and Foundations: Geotechnical Engineering*, 3d ed., Macmillan, New York, 1970.

Sowers, G. F., *Introduction to Soil Mechanics and Foundations: Geotechnical Engineering*, 4th ed., Macmillan, New York, 1979.

Sponenberg, T. D., et al., "A Homeowners Guide to Septic Systems," Virginia Water Resources Research Center, VPI, Blacksburg, VA, 1985.

Truitt, M. M., *Soil Mechanics Technology*, Prentice-Hall, Englewood Cliffs, NJ, 1983.

Upton, R. J. and G. Machan, P.E., "Use of Shredded Tires for Lightweight Fill," Transportation Research Board, 72nd Annual Meeting, Washington, DC, 10–14 January 1993.

U.S. Environmental Protection Agency, *Hazardous Waste Sites: National Priorities List*, HW-7.1, August 1983.

Wager, O. and R. D. Holtz, "Reinforcing Embankments by Short Sheet Piles and Tie Rods," *New Horizons in Construction Materials*, Vol. I, Envo Publ., Lehigh Valley, PA, 1976.

Wetlands Ecology and Conservation: Emphasis in Pennsylvania, Pennsylvania Academy of Science, Easton, PA, 1989.

"Wetlands: Untangling the Issues and Regulations," Proceedings, WRA/DRB 30th Annual Spring Conference, 13 April, 1989, Cherry Hill, NJ, WRA/DRB, Valley Forge, PA.

Wilcox, K., "Little Common Ground in Septic Additive Debate," *Small Flows*, July 1992, p. 1.

Williams, W., "Development and Use of Plastic Foam Insulation to Prevent Frost Action Damage to Highways—A Summary of Experience in the United States," *International Conference on Highway Insulation*, Wurzburg, Germany, May 1968.

Yeh, S. T. and J. B. Gilmore, "Application of EPS for Slide Correction." Stability and Performance of Slopes and Embankments-II, R. B. Seed and R. W. Boulanger (eds.), ASCE, 1992, pp. 1444–1456.

Yoder, E. J., *Principles of Pavement Design*, Wiley, New York, 1959.

Zwingle, C., "On Specifying Percentage Compaction for Clay Soils," Master's project (unpublished), New Jersey Institute of Technology, Newark, NJ, 1981.

INDEX

AASHTO system, 8
Absorption field, 195
Acceptable risks, 152
Acquisition, of borrow, 130
Aeration, 112
Aerobic systems, 194, 224
  Cromaglass, 225
  Rotordisk, 240
Air-drying (laboratory compaction), 23
Air permeability, 103
Air photos, 130
Airport construction, 63
Allowable bearing pressure, 33
Ambiguity, avoidance, 161
Angularity, 51, 97, 252
Anistropy, 49, 94
Approach fills, 55, 61
Architect, 2, 129
Arrow hammer, 110
Artificial fills, 53
  case histories, 56, 72
  Elastizell, 72
  expanded polystyrene (EPS), 71
  future usage, 69
  Geocell, 73
  history (of development), 69
  hybrid design, 73
  patents, 70
  research, 71
  Solite, 73
  Styrofoam HI, 54
Artificial subbase, 63
Ash, fills, 74
ASTM Compaction Requirements, 22
ASTM designations, 144
Atterberg Limits, 8, 130

"Backfill Guide," 108
Bad weather, 116
Ballast, 111
Bamboo, 74
Bay muds, 102
Bearing capacity, 135
  allowable, 2, 3
  presumptive, 139
Bedrock, depth, 130
Blasting, 131
Blister densities, 145
Blow counts, 4, 6
Blown-in-place foam fills, 57, 64
Blunders, avoidance, 12, 16
Boiling, 87
Boring(s), 117
Boring depths, 135
Boring inspection, 157
Boring program, 138
Borrow:
  area, mound, 118
  changing, 40, 103, 120
  heterogeneous, 118
  loading equipment, 131
  sources, 130
  texture, changes, 120
Breaching, 91, 92

Bridging, in fills, 103
Building codes, 132
Burmister system, 8, 212, 220

California Bearing Ratio, 115, 122
Case histories, 166
Caving, 87
Certification, 118, 128
  levels of, 154
Changing borrow, 40, 103, 120
Cheap jobs, 151, 152
Chimney drain, 97, 228
Chunk samples, 145
Chunk-volume determinator, 148
Cinder fills, 62
Classification:
  plastic soils, 20
  of soils, 104
Classification system(s), 8, 211
  auxiliary, 5
Clay(s), lean, fat, 20
Clay fill, 117
Client, 128, 129
Coefficient of permeability, 87
Cohesionless soils, 3
Cohesive soils, 3
Cold weather, 121
Cold weather gear, 157
Collapse, of clays, 108
Communication with clients, 154
Compacted clays, 94
Compacted density, condition, 21
Compaction:
  blunders, 12
  definition, 17
  field, 3
  granular soils, 124
  history, 12
  laboratory, 3
  modified Proctor, 14
  moisture content, 18, 25
  standard and modified Proctor, 15
  standard Proctor, 13
Compaction control, 119
  CBR, 122
  comprimeter, 125
  degree of compaction, 119, 122
  field density testing, 126
  by index properties, 125
  nuclear method, 126
  plate loading, 123
  pocket penetrometer, 125
  proof rolling, 126, 132
  special circumstances, 126
  water balloon, 126
Compaction curves, locations, shapes, 20
Compaction Data Book, 40, 47
  corrections, 50
Compaction equipment, 132
Compactor(s), 103
  hand, 109
  types, 104
Compleat field inspector, 156
Completion time, 129
Confined spaces, 65, 106, 122, 141
Consequences, distress, failure, 151
Consistency ratings, clays, 217
Construction engineer, 129
Construction rubble, 73
Contact pressure, 135
Contract documents, 129
Contractor, 129, 181
Core wall, 86
Coring operations, 177, 252
Correction, 129
Cost:
  factors, 151
  pressures, 45
  Styrofoam, 55
Critical exit hydraulic gradient, 91
Cut-and-cover construction, 62
Cutoff wall, 86

D_{10} size, 11, 218
Dense sand, 108
Density corrections, 50
Density gradients, 81
Density requirements, 130
Density test coordinates, 178
Density testing, how much?, 149
Development, research, 47
Differential icing, of pavements, 63, 67
Differential settlements, 134
Dilatancy test, 10, 215
Direction measurements, 157
Disclaimers, 155
Distance measurements, 157
Drainage characteristics, 130
Drainage designations, 4
Drainage (drain) field, 193
Drum rollers, 104
Dry borrow, 121
Dry density, 19
Drying, scarification, 120
Dry of optimum, ,83, 95
Dumped backfills, 106

Earth dam(s), 2, 86, 117
  caveat, 84
Earth structures, 87, 92

Earthwork contractors, 169
Earthwork equipment, 131
Elephants, 106
Elevation measurements, 157
Embankment, 95
  stability, 99
End-result specifications, 26, 115, 119
End-result and suggested method, 119
Energy (of) compaction, 110
Energy control, 110
Engineering properties, 2, 123
  as design basis, 151
Engineering technologists, 128
Environmental factors, 152, 190, 217
Environmental Protection Agency, 102
Equipment imbalance, 168
Erosion, 94
Ethics, 28
Euclid truck(s), 105
  as compactors, 170
Excavated materials, 117
Excavated surfaces, 117
Excavating, 117
Excavation, ease of, 131
Excavation equipment, 131
Exclusion, in laboratory compaction test, 23
Exculpatory clauses, 156
Expansiveness, 84, 86
Expensive jobs, 151
Expert witness, 127

Fat clays, 49
Favors, gratuities, 185
Floating foundation, 62
Flocculent structure, 95
Flood control dams, 99
Flooding, 109
Flotation, 68
Field clothing, 157
Field density checks, 120
Field density hole, 141
Field density techniques, 120
Field density tests, 14, 141, 173
  locations, 120
Field equipment lists, 157
Field moisture, 116
Field notes, 162
Field observations, 176
Field testing costs, 151
Field tests, soil texture, 214
Fill, 5, 101
  condition, 101, 117
  controlled, 101
  correction measures, 121
  load-bearing, 101
  partially controlled, 101
  placement, 118
  special, 101
  texture, 101, 117
  uncontrolled, 101
  undulation, 17, 24
  use of, 117
  waste, 73
Fill compaction, 86
Fill control procedures, 134
Fill inspector, 44
Fill records, 176
Filmstrips, 251
Fines, 117
Foam plastic:
  fills, 53
  poured, 64
  rigid, 64
Foundation:
  design, 138
  failure, 127
  selection, 138
  suitability, 132
Four-year fill option, 224
Frost action, 87, 92
Frost-susceptible soils, 92
Frozen soil, 121

Garbage, 74
Garbage In, Garbage Out, 156
Geodex, 105
Geostick, 81, 139
Geotechnical engineer, 2
Glacial till, 131
Glass, as fill, 74
Graded filter, 97
Grade-separation case study, 61
Grain size distribution, 214, 218
Grain size limitations, 117

Hand sieving, 163
Haul distance, 130
Hauling equipment, 131
Hazardous fills, 153
Hazen's effective size, 50
Heel test, 140
Highway(s), 58
  insulation, 56
Honeycomb structure, 95
Horizontality measurements, 157
Hydraulic fills, 101
Hypothetical case histories, 57

Index properties, 2, 123
  as design basis, 151

Index property variations, 42
Infiltrator system, 209, 223
Infrastructure repair, 93
In situ, 46
Inspection, 128, 134
Inspector, 128, 134, 140, 143, 154, 181, 184
Instant moisture balance, use, 24
Insulation, 56, 77

Jacked sample method, 145
Jetting, 109

Laboratory procedures, 251
Laboratory techniques, 220, 251, 252
Labor costs, 106
Laws and regulations, 132
Lawsuits, 127
Leachate control, 102
Leach field, 193
Lean clays, 49
Legal costs, 243
Level adjustments, 164
Liability insurance, 243
Lift preparation, 118
Lift thickness,115, 118
Lightweight fills, 53
Lime-stabilized soils, 122
Limitations, reasonable, 155
Limiting zone, 193, 195
Liquefaction, 92
Liquid limit, 125
Litigious society, 126
Load-bearing fills, 116
Loaded area shapes, 135
Load testing, 163
  setup, 164
Long-term compression, 68
Loose sand, 108, 125
Lunch hour filling, 122

Major problems, 26
  ninety-five percent fixation, 29
  standard-modified ignorance, 26
Making do, 105
Marginally permeable, 87
Masking (of drain field), 209
Mechanical compactors, 80
Methods specifications, 115
Mineralogy, 117
Moisture computation, 111
Moisture content, 118
Moisture control, 110
  adjustments, 118
Moisture and energy sensitivity, 21, 25
Moisture sensitivity, 49, 99
Mottling, 195

National Dam Inspection Program, 87
National Sanitation Foundation, 225
Non-load-bearing fills, 116
Nonplastic, fines, 48
Nonspecialists, 1, 2, 43, 114
Nuclear density method, 148

Observer, 128
Odors, 102
Optimum moisture content, 13, 21, 24
  corrections, 50
Oral records, 178
Organics, 103
OSHA, 132
Ottawa sand, 125, 143
Ozone depletion, 68

Pace length, 160
Passes, number of, 119
Payment, terms, 129
Percentage compaction, 119
  for clays, 83
  example, 36
  method of determining, 32
Percent fines, 41, 48
Percent saturation, 25
Perched water table, 197
Percolation testing, 193, 203
Permanency, 55
Permeability, 50
Personal safety, 186
Photo records, 178
PI, of fines, 41, 48, 214
Pickford Bridge, 54, 65
Pile load, test, 60
Pipelines, 59
Piping, 91
Placement equipment, 131
Plastic, fines, 49
Plasticity, 214
Plasticity index, 117
Plate bearing tests, 115, 123, 140
Plate loading test, field, 123
Plumb measurements, 157
Pneumatic rollers, 104
Pocket penetrometer, 139
Poleset, 64
Postcompaction precautions, 121
Precast foam fills, 53
Precompaction equipment, 131
Pressure bulb, 135
  uses, exceptions, limitations, 137

Presumptive bearing capacity, 81
Proctor density, 122
  bearing capacities, 38
  correlations, 38
Professional liability insurance, 127
Proof rolling, 132
Pumping, 93

Qualitative fill control, 179
Quantity, of borrow, 130
Quartering, soil samples, 25
Quick condition, 87

Rain problems, 120, 121
Random selection technique, 179
Range of sizes, C_R, 218
Rapid drawdown, 99
Recompaction effects, 22
Reconnaissance, 156
Record keeping, 157, 180
Reinforced earth, 63
Relative density, 34, 97, 107
Report writing, 176
Representative batches, preparation, 24
Representative samples, 118, 173
Research, 47
Retaining structures, 63, 108
Richter scale, 11
Road bases, 115, 122
Roadway embankments, 116
Rock, 97
  estimate, quantity, 131
  rippability chart, 131
  ripping, blasting, 131, 187
Roofing, 87
Route construction, 93
Rubber, as fill, 73, 74
Rubble fills, 103

Sample bags, 162
Sample disturbance, 148
Sampling ratio, 144
Sand cone apparatus, 142
Sand cone method, 141
Sand cone test, steps, 143
Sanitary fills, 102
Scarification, drying, 118
Scraper, 131, 169
Seepage, 90
  losses, 90
  pressures, 90
Select fill, 10, 22, 25, 96, 117. *See also* Septic systems
Seminars available, 220, 252
Septic systems:
  additives (chemical, biological), 237
  aerobic systems, 224
  alternating systems, 191, 229
  background technology, 191
  bed systems, 192, 206
  Cook County Cottege, 211
  Department of Environmental Resources, 194, 202, 230, 240
  design/construction, 206
  dribble stone, 219, 228
  environmental regulations, 191, 194, 195, 197, 222
  experimental systems, 194
  filter fabric, geotextiles, 208, 209
  harsh chemicals, 231
  honey wagons, 235
  maintenance, 191, 195, 230
  management, 240
  mound systems, 193, 210
  oil-eating bacteria (DBC-Plus), 238
  pollution science, 191
  pressure distribution system, 211, 227
  problem sites, 222
  renovation/repair, 229
  Rotordisk system, 224
  seasonal high water table, 195
  seepage analysis, 191
  select fill, 211
  septic tank no-no's, 231
  septic tank pumping, 235, 236
  site investigation/evaluation, 195
  standard systems, 191, 206
  suspended solids, 207, 231
  Tabasco sauce(!), 223
  test pits (probes), 193, 195
  Thetford system, 226
  trench backfill, 209
  trench preparation, 208
  trench systems, 192, 206
  water conservation, 232
Settlement(s), 86
  computation, 107
  plates, 120
Shear failure, 99
Sheepsfoot roller, 104
Shelby tube sampling, cohesive soils, 217
Shell, of dam, 86
Shocks, 92
Silts, 21, 93, 98
Simulation, of field conditions, 110, 142
Sliding, 99
Slope stability, 99
Soft wet clay, 108
Soil classification, visual, 163, 215
Soil Conservation Service (SCS), 130

Soil identification, classification, 189, 214
Soil maps, 130
Soil sampling, 252
Specialist's services, 118, 131
Specifications, 3, 114
  authority, 128
  careless, 31
  enforcement, 126, 185
  evaluation, 129
  implementation, 126
  nontechnical aspects, 129
  responsibility, 128
  special criteria, 122, 126
Specific gravity, 19
Split spoon sampling, granular soils, 217
Splitting, soil samples, 25
Sprinkling, field, 112, 118, 132, 170
Stability, 5, 99
  of excavations, 181
Standard Penetration Test, 32, 46, 216
Static loadings, 125
Stripping and grubbing, 27, 117, 130
Structural engineer, 2, 114, 129, 134
Structural fill, 108, 116
Structure of clays, 94
Styrofoam HI, 54
Subbases, 122
Subgrade, 95, 115, 137
Subgrade modulus, 123
Subgrade stabilization, 58
Subsidence, 175
Suggested guide specifications, 116
Suggested method and end result, 115
Supercompactors, 105
Surface damage, 121
Surveying techniques, 161

TALB, 20
Tamping rollers, 104
Target value density, 26, 115, 141, 173
Techonomics, 242
Test pits, 117
  locations, 179
  sampling, 162
Test strips, 119
Textural tests, 145
Texture, 20, 117
  acceptable, 126
  conformance, 142
  of borrow, 130
Time, pressures, 45
Total cost context, 57
Traditional practice (causing problems), 43
Traffic compaction, 208
Trafficking problems, 130
Traffic vibrations, 107
Trench backfill, 106

Unacceptable end results, 129
Uncompacted fills (densities), 81
Unconfined compressive strength, 125, 139, 217
Undulation of fill, 120
Unified Soil Classification System, 104
Uniformity coefficient, 46, 51, 223
Unsuitable material, 117
Uplift pressure, 85
Utility installations, 56

Varved clays, 136, 189
Vector control, 102
Vegetation cover, 102
Venting, gases, 102
Verification, degree of, 153
Very large jobs, 247
Vibrations, 124
  postconstruction, 125
Vibratory rollers, 104
Vibroflotation, 105
Void ratio, 46, 107

Washington Densometer, 144
Waste fills, 73
  case stsudies, 73
  energy conservation proposal, 76
  environmental regulations, 75
  geotextile reinforcement, 75
  rabbits (!), 78
  research, 75
  rubber plastic, 78
  steel slag, 77
  telephone books, 77
  tire residue, 77
  tires (shredded), 75
Waste materials, 73
Water absorption, 55
Water balloon method, 144
Water table, depth, 130
Weather restrictions, 120
Weep holes, 109
Weight credit, 110
Wet borrow, 120
Wetlands, 194, 199

buffers, 200
categories, 200
definition/delineation, 199
policy, 201
Wet of optimum, 83, 95
Winneberger "perk" test, 204, 207
Work changes, 129
Written criticism, 180